SpringerBriefs in Physics

SpringerBriefs in Physics are a series of slim high-quality publications encompassing the entire spectrum of physics. Manuscripts for SpringerBriefs in Physics will be evaluated by Springer and by members of the Editorial Board. Proposals and other communication should be sent to your Publishing Editors at Springer.

Featuring compact volumes of 50 to 125 pages (approximately 20,000–45,000 words), Briefs are shorter than a conventional book but longer than a journal article. Thus, Briefs serve as timely, concise tools for students, researchers, and professionals.

Typical texts for publication might include:

- A snapshot review of the current state of a hot or emerging field
- A concise introduction to core concepts that students must understand in order to make independent contributions
- An extended research report giving more details and discussion than is possible in a conventional journal article
- A manual describing underlying principles and best practices for an experimental technique
- An essay exploring new ideas within physics, related philosophical issues, or broader topics such as science and society

Briefs allow authors to present their ideas and readers to absorb them with minimal time investment. Briefs will be published as part of Springer's eBook collection, with millions of users worldwide. In addition, they will be available, just like other books, for individual print and electronic purchase. Briefs are characterized by fast, global electronic dissemination, straightforward publishing agreements, easy-to-use manuscript preparation and formatting guidelines, and expedited production schedules. We aim for publication 8–12 weeks after acceptance.

More information about this series at http://www.springer.com/series/8902

Michael Charlton · Stefan Eriksson ·
Graham M. Shore

Antihydrogen and Fundamental Physics

 Springer

Michael Charlton
Department of Physics
College of Science
Swansea University
Swansea, UK

Stefan Eriksson
Department of Physics
College of Science
Swansea University
Swansea, UK

Graham M. Shore
Department of Physics
College of Science
Swansea University
Swansea, UK

ISSN 2191-5423   ISSN 2191-5431   (electronic)
SpringerBriefs in Physics
ISBN 978-3-030-51712-0   ISBN 978-3-030-51713-7   (eBook)
https://doi.org/10.1007/978-3-030-51713-7

This Springer imprint is published by the registered company Springer Nature Switzerland AG
The registered company address is: Gewerbestrasse 11, 6330 Cham, Switzerland

Acknowledgements

MC and SE are members of the ALPHA collaboration. Their work is supported by EPSRC grant EP/P024734/1. The work of GMS is supported in part by the STFC theoretical particle physics grant ST/P00055X/1. We would like to thank M. Fujiwara, N. Madsen, F. Robicheaux, S. Jonsell and D. P. van der Werf for helpful discussions and C. Ø. Rasmussen for valuable comments on the manuscript.

Contents

Chapter 1
Introduction

The recent measurement by the ALPHA collaboration of the $1S$–$2S$ spectral line in antihydrogen with a precision of a couple of parts in 10^{12} [1, 2] marks the beginning of a new era of precision anti-atomic physics. Future experiments on antihydrogen and other antimatter species will enable exceptionally high-precision tests of many of the fundamental tenets of relativistic quantum field theory and general relativity, such as CPT invariance, Lorentz symmetry and the Equivalence Principle. It is therefore timely to examine critically what each of these experiments may be said to test and what any violation from standard expectations would mean for fundamental physics.

Experiments on pure antimatter systems, whether elementary particles or bound states such as antihydrogen, are especially interesting from this point of view since they are constrained so directly by the fundamental principles underlying the standard model. For example, the discovery of a new Z' boson, right-handed neutrinos, a supersymmetric dark matter candidate etc. would be of immense interest but could readily be assimilated into an extension of the standard model. In contrast, an anomalous result on the charge neutrality of antihydrogen, or a difference in the $1S$–$2S$ transitions of hydrogen and antihydrogen, would impact directly on the foundations of local relativistic QFT. In these theories, the existence of antiparticles with precisely the mass and spin, and opposite charge, of the corresponding particles is required by Lorentz invariance and causality. Moreover, for a *local* QFT, Lorentz invariance implies invariance under CPT, according to the celebrated theorem [3–6]. Antimatter experiments therefore directly test these principles.

The situation is not so clear when we consider gravity, where such experiments are often presented as tests of "the equivalence principle". The difficulty is that there are several versions of the equivalence principle in the literature—weak, strong, Einstein—with definitions which are not always either unique or well-defined. Indeed, as emphasised by Damour [7], it should not really be considered as a 'principle' of GR in the more rigorous sense that Lorentz symmetry and causality are principles of QFT. A more satisfactory approach is to recognise that we have a well-defined, and extraordinarily successful, theory of gravity in GR, which makes clear and precise predictions for the gravitational interactions of all forms of matter. Like other experiments, those on antimatter should simply be viewed as tests of this theory.

General Relativity is based on the idea that gravitational interactions may be described in terms of a curved spacetime. As described more precisely in Sect. 2.4, this spacetime is taken to be a Riemannian manifold, since this has the property that at each point it locally resembles Minkowski spacetime. The global Lorentz symmetry of non-gravitational physics is reduced to *local Lorentz symmetry* in curved spacetime. This is the mathematical realisation of the physical requirement of the existence of local inertial frames (i.e. freely-falling frames) even in the presence of gravity. Further to this, the standard formulation of GR makes a simplifying, though well-motivated, choice of dynamics for the interaction of matter and gravity, which is encapsulated in the following statement of the Strong Equivalence Principle:

- *In a local inertial frame, the laws of physics take their special relativistic form* (SEP).

  We will also discuss frequently two further expressions of the universality at the heart of GR. These are best viewed as experimental predictions of GR, though we refer to them here as versions of the Weak Equivalence Principle:

- *Universality of free-fall—all particles (or antiparticles) fall with the same acceleration in a gravitational field* (WEPff).
- *Universality of clocks—all dynamical systems which can be viewed as clocks, e.g. atomic or anti-atomic transition frequencies, measure the same gravitational time dilation independently of their composition* (WEPc).

Taken together, these three properties of GR are usually referred to as the *Einstein Equivalence Principle.*[1]

Apparent violations of these predictions, especially WEPff, can also arise not from the actual violation of any fundamental principle of QFT or GR but from the existence of new interactions not present in the standard model, so-called 'fifth forces'. Low-energy precision experiments on antimatter, whether involving spectroscopy or free-fall equivalence principle tests, may be sensitive to such new interactions and can place limits on their range and coupling strength. Here, we consider two such possibilities, both well-motivated by fundamental theory. The first is an extension of the standard model gauge group to include a new $U(1)_{B-L}$ factor, with a corresponding gauge boson Z' coupling to $B - L$ (baryon minus lepton number) charge. The second involves the spin 1 'gravivector' boson which arises in some supergravity

---

[1]The Einstein Equivalence Principle may be stated in various essentially equivalent ways. In [8], the three principles are referred to as *Local Lorentz Invariance* (LLI), which we have called SEP; the *Weak Equivalence Principle* (WEP), which is simply our WEPff; and *Local Position Invariance* (LPI), which states that 'the outcome of any local non-gravitational experiment is independent of where and when in the universe it is performed' [8]. LPI implies the universality of gravitational redshift, or WEPc, and can also be tested through the space and time-independence of fundamental constants. Note that while GR implies WEPc, the latter is a more general property of any metric theory of spacetime. Also note that, as described above, a metric theory like GR on a Riemannian spacetime manifold exhibits LLI, but the dynamics need not be the same as special relativity, or be independent of the local curvature, if SEP is violated.

theories with extended, $\mathcal{N} \geq 2$, supersymmetry. Both have the potential to modify gravitational free-fall in violation of WEPff, distinguishing between matter and antimatter. We also consider a more general phenomenological approach to the possible existence of new, gravitational strength, vector or scalar interactions.

From an experimental perspective, the study of the fundamental properties of antiparticles and atomic systems constituted wholly, or partially, from them is a growing area of endeavour. In this book, our main focus is on antihydrogen, and in particular the current experiments being performed by the ALPHA collaboration at CERN and their implications as tests of fundamental physics. Later, we briefly consider a range of other antimatter species which may offer complementary opportunities for such tests.

We start by describing some of the practical aspects of current experiments with $\overline{H}$ at low energy. Positrons (e^+) are available in the laboratory, typically via pair production and from radioactive materials (see e.g. [9] for a review). We concentrate on the latter, and the isotope ^{22}Na (half-life around 2.6 years, β^+ fraction about 90%) is the typical choice of source. Sealed capsules of around GBq activity can be held in vacuum and, using well-documented procedures (see e.g. [10]), eV-energy beams can be produced with efficiencies of 0.1–1% of the source strength. Such beams can be readily transported in vacuum to devices which enable their trapping and accumulation: the most common instrument to achieve this is the so-called buffer gas trap which, using a Penning-type trap and energy loss via inelastic positron-gas collisions, can accumulate around 10^8 e^+ in a few minutes, if required. The positrons can then be transferred [11] on demand for further experimentation, and of most relevance here for $\overline{H}$ production and trapping.

Antiprotons ($\overline{p}$) are only available at laboratories such as CERN where high energy protons (typically 20–30 GeV) produce the $\overline{p}$s in collision with fixed targets. CERN's unique Antiproton Decelerator (AD) [12, 13] syphons off $\overline{p}$s at a kinetic energy of about 3.5 GeV and then decelerates and cools them in stages to reach 5.3 MeV, whereupon they are ejected to experiments in bursts of 100 ns duration containing around 10^7 particles, about once every 100 s. The kinetic energy of the $\overline{p}$s is typically moderated using foils whose thickness is carefully adjusted to maximise the transmitted yield (of around $10^{-2} - 10^{-3}$ of the incident flux) below 5–10 keV, and these are then captured in dynamically switched high field Penning traps [14] where they can be efficiently electron cooled [15, 16] to sub-eV energies. The $\overline{p}$s and electrons can then easily be separated, and the former then transferred to another apparatus or stored for further manipulation and experimentation.

The mixing of $\overline{p}$s and e^+s to form $\overline{H}$ has been described in detail elsewhere [17, 18], and a number of techniques have been developed to hold the antiparticle species in close proximity and manipulate the properties of the respective clouds (e.g. number, density and temperature) in a system of Penning traps to facilitate anti-atom creation. Under the conditions of e^+ cloud density (around 10^{14} m^{-3}) and temperature (typically in the range 5–20 K) commonly used in $\overline{H}$ experiments, the dominant formation reaction is the three body process $e^+ + e^+ + \overline{p} \rightarrow \overline{H} + e^+$. It is well-documented (see e.g. [19, 20]) that this reaction produces highly excited $\overline{H}$ states: thus, if experimentation on the ground state is required, the neutral should be held in a suitable trap, otherwise most of the nascent anti-atoms will annihilate on contact with the Penning trap walls, typically within a few μs.

The traps used for antihydrogen have, to date, been variants of the magnetic minimum neutral atom trap [21, 22], as widely applied in cold atom physics (see e.g. [23]). The trapping force is due to a magnetic field gradient acting upon the anti-atom magnetic moment, such that the low-field seeking anti-atoms (i.e. these whose positron spin is anti-parallel to the local magnetic field) are trapped. The practical details need not concern us here (see [24]). However, even with advanced superconducting magnet technology the traps are shallow, at around 0.5 K deep, with respect to the aforementioned $\overline{p}/e^+$ temperatures on mixing. Thus, only a small fraction of the antihydrogen yield can be captured [18, 21, 26, 27], and the recent state-of-the-art is about 10–20 trapped anti-atoms from around 50,000 created during mixing of 90,000 $\overline{p}$s with 3 million e^+ at around 15–20 K [28]. Nevertheless, the very long lifetime of the antihydrogen atoms in the cryogenic trap environment—in excess of 60 h [2]—has allowed their accumulation over extended periods [18] with the record being over 1500 $\overline{H}$s in ALPHA's 400 cm^3 magnetic trap. It is from this basis, of $1 - 10^3$ stored $\overline{H}$s that the experiments described below and in Chap. 3 have been achieved.[2]

So far, all precision experiments with antihydrogen in the ALPHA apparatus rely on detecting the by-products of annihilation when the anti-atom escapes from the magnetic trap and comes in contact with the (matter) walls of the electrodes used to confine the charged particle plasmas during antihydrogen synthesis. The high-energy particles (mostly pions) produce hits in silicon strips arranged axially in three cylindrically shaped layers surrounding the exterior of the trapping apparatus [25]. The particle tracks are then traced back and the location of the annihilation event is found by searching for a vertex occurring on the electrode wall. Artificial learning is used to efficiently distinguish between background events and as a result, the tracking detector is able to spatially resolve single antihydrogen annihilation events. Antihydrogen detection events in this silicon vertex detector occur in two modes. In disappearance mode an experiment which produces loss of anti-atoms from the trap, e.g. by resonantly inducing a transition to an untrapped state, is conducted, after which the trap is rapidly turned off and the remaining anti-atoms are detected. In appearance mode anti-atoms escaping due to the interaction are detected while the trap is energised. The modes of detection can be applied together, allowing consistent monitoring of the trapping rate.

The first precision experiment carried out by ALPHA to interrogate the properties of antihydrogen was the verification of charge neutrality [30, 31]. Interpreted as confirmation that the electron and positron, and the proton and antiproton, have equal

---

[2]The antihydrogen experiments are clearly $\overline{p}$ flux limited and in promotion of this field CERN is developing ELENA, an ultra-low energy add-on to the AD [29]. This will provide $\overline{p}$s at 100 keV (rather than 5 MeV), which will enhance capture efficiencies in most experiments by a factor of around 100. Furthermore, the low $\overline{p}$ kinetic energy will allow delivery using electrostatic transport and switching technology, thus facilitating rapid changeover of beam between experiments. This will result in a $\overline{p}$ on-demand mode of operation, which together with the higher capture efficiencies will vastly enhance capabilities.

and opposite charges, it is clearly a requirement of CPT invariance. We emphasise, however, that this basic property of antiparticles is much more fundamental and is necessary to preserve causality. The exact equality of the magnitudes of the proton and electron charges is in turn necessary in the standard model to ensure unitarity.

Control over the dynamics of the antihydrogen atoms in the ALPHA magnetic trap [32, 33] allowed a preliminary gravity experiment [34] setting the loose bound $F \lesssim 110$ on the ratio $F = m_g/m_i$ of the 'gravitational mass' to the 'inertial mass'. These definitions are discussed critically here in Sects. 2.4 and 3.3. Far more precise bounds on WEPff violation for antimatter, at the percent level, are expected with the dedicated ALPHA-g apparatus and other experiments in development: see Sect. 3.3.

The spectroscopy programme, which we discuss in more detail in Chap. 3, began with a demonstration of microwave-induced spin flips in antihydrogen [35], and further development has resulted in the current high-precision measurement of the hyperfine structure. A measurement of the hyperfine splitting for antihydrogen from the difference of the $1S_d - 1S_a$ and $1S_c - 1S_b$ transitions was presented in [36] and found to be in agreement with hydrogen at the 10^{-4} level, consistent with expectations.

The gold standard is of course the two-photon $1S-2S$ transition, which for hydrogen has been experimentally determined with a frequency precision of 4.2×10^{-15} [37]. In [1], ALPHA verified this result for antihydrogen with a precision of 10^{-10}. Note that the anti-atom measurement is made on the hyperfine state transitions $1S_d - 2S_d$ and $1S_c - 2S_c$ in the ~ 1 T magnetic field of the confining trap. More recently [2], this precision has been improved still further with the measurement of the antihydrogen $1S-2S$ transition at the level of 2×10^{-12}. In terms of relative precision of the experimental measurement, this is the most precise test of CPT symmetry achieved to date with the anti-atom.

So far, therefore, the ALPHA collaboration investigations of antihydrogen have proved consistent with the fundamental principles outlined above, especially CPT. This is, however, still only the early stages of an extensive programme of high precision spectroscopy [38]. Gravity tests, of both WEPff and WEPc, as well as explicit tests of Lorentz symmetry, are just beginning and will achieve competitive levels of precision in the coming years, exploiting the special nature of antihydrogen as a neutral, pure antimatter state [38]. Atom interference experiments with antihydrogen are also feasible, while ultra-high precision spectroscopy with molecular $\overline{H}$ states can be envisaged.

The book is organised as follows. Chapter 2 describes the main fundamental principles which will be tested in these experiments, especially Lorentz, CPT, causality, unitarity and the equivalence principles embodied in GR. Modifications to standard theory are considered, ranging from effective field theories incorporating Lorentz, CPT or SEP breaking to models with new 'fifth forces' which would modify WEPff especially.

Chapter 3 is devoted to antihydrogen. We describe the ALPHA spectroscopy programme and derive explicit formulae for the dependence on the Lorentz and CPT violating couplings in the SME effective theory [39, 40] for the specific transitions between $1S$, $2S$ and $2P$ hyperfine states measured by ALPHA. We then turn to gravity and, after a review of the planned experimental programme for antimatter equivalence principle tests with ALPHA-g, GBAR and AEgIS at the CERN AD, we discuss how these are interpreted in the framework of GR and possible EP-violating phenomenological models. A careful discussion of physical time measurement in the context of metric theories such as GR is given in Chap. 2 as background to the interpretation of these experiments.

Finally, in Chap. 4, we briefly describe a number of other antimatter bound states which could be studied experimentally in future as complementary tests of fundamental physics principles. A general summary and outlook is presented in Chap. 5.

Chapter 2
Fundamental Principles

We begin by reviewing some of the fundamental principles of local, relativistic QFT and GR that may be tested in current and forthcoming low-energy experiments on antimatter, especially with antihydrogen. The discussion is elementary and is intended simply to highlight how basic principles such as Lorentz invariance, CPT, the equivalence principle, etc. are embedded in the standard theories of particle physics and gravity.

2.1 Antimatter and Causality

The first key question to address is why antiparticles exist and why their properties (mass, spin, charge ...) must *exactly* match those of the corresponding particles. While this is often presented in historical terms, going back to early and now superseded interpretations of the Dirac equation in relativistic quantum mechanics, the real reasons for the existence of antiparticles are much deeper and more general, and certainly not specific to spin 1/2 particles. In fact, antiparticles are required to maintain causality in a Lorentz invariant quantum theory.

First, we recognise that single-particle relativistic quantum mechanics is not a consistent theory and has to be replaced by a quantum field theory, with the Dirac equation promoted to a field equation for a spinor quantum field $\psi(x)$. The corresponding action for QED, with $\psi(x)$ representing the electron and the gauge field $A_\mu(x)$ describing the photon, is

$$S = \int d^4x \, \mathcal{L}_{\text{QED}} = \int d^4x \left(\bar{\psi} \left(i\gamma^\mu D_\mu - m \right) \psi - \frac{1}{4} F_{\mu\nu} F^{\mu\nu} \right), \quad (2.1)$$

with $D_\mu = \partial_\mu - ieA_\mu$.

In this Lorentz-invariant QFT, the Dirac field $\psi(x)$ is expanded (in the Heisenberg picture) in terms of creation and annihilation operators for electrons and positrons as

© The Author(s), under exclusive license to Springer Nature Switzerland AG 2020
M. Charlton et al., *Antihydrogen and Fundamental Physics*,
SpringerBriefs in Physics, https://doi.org/10.1007/978-3-030-51713-7_2

$$\psi_\alpha(x) = \int \frac{d^3\mathbf{p}}{(2\pi)^3} \frac{1}{\sqrt{2E_\mathbf{p}}} \sum_s \left(a_\mathbf{p}^s u_\alpha^s(p)e^{-ip\cdot x} + b_\mathbf{p}^{s\dagger} v_\alpha^s(p)e^{ip\cdot x}\right), \qquad (2.2)$$

where p^0 is identified as $E_\mathbf{p} = \sqrt{\mathbf{p}^2 + m^2}$, according to the standard relativistic dispersion relation, and the momentum integral is over the Lorentz invariant measure. The adjoint admits a similar decomposition with $a_\mathbf{p}^s \leftrightarrow b_\mathbf{p}^s$ and $u^s(p) \leftrightarrow \bar{v}^s(p)$. Note that these forms assume the full symmetries of Minkowski spacetime, including translation invariance. The charge conjugate field $\psi^C(x) = C\bar{\psi}^T(x)$, with $C = i\gamma^0\gamma^2$, takes the form

$$\psi^C(x) = \int \frac{d^3\mathbf{p}}{(2\pi)^3} \frac{1}{\sqrt{2E_\mathbf{p}}} \sum_s \left(b_\mathbf{p}^s u_\alpha^s(p)e^{-ip\cdot x} + a_\mathbf{p}^{s\dagger} v_\alpha^s(p)e^{ip\cdot x}\right). \qquad (2.3)$$

Here, $a_\mathbf{p}^s$ ($a_\mathbf{p}^{s\dagger}$) is the annihilation (creation) operator for an electron of spin s and momentum $\mathbf{p}$, while $b_\mathbf{p}^s$ ($b_\mathbf{p}^{s\dagger}$) are the corresponding operators for positrons. Under charge conjugation, $a_\mathbf{p}^s \overset{C}{\to} b_\mathbf{p}^s$. The spinors $u_\alpha^s(p)$, $v_\alpha^s(p)$ are the standard solutions of the Dirac equation, satisfying notably[1]

$$\sum_s u_\alpha^s(p)\bar{u}_\beta^s(p) = (\gamma\cdot p + m)_{\alpha\beta},$$

$$\sum_s v_\alpha^s(p)\bar{v}_\beta^s(p) = (\gamma\cdot p - m)_{\alpha\beta}. \qquad (2.4)$$

The free Dirac theory has a global $U(1)$ symmetry, promoted to a local $U(1)$ with the inclusion of the photon field $A_\mu(x)$ in QED, with Noether current $J^\mu = \bar{\psi}\gamma^\mu\psi$. This implies the existence of a conserved charge, the electric charge in QED, which is expressed in terms of the number operators for electrons and positrons as

$$Q = \int d^3x \, J^0 = \int \frac{d^3\mathbf{p}}{(2\pi)^3} \sum_s \left(a_\mathbf{p}^{s\dagger}a_\mathbf{p}^s - b_\mathbf{p}^{s\dagger}b_\mathbf{p}^s\right). \qquad (2.5)$$

This shows that the antiparticles appearing in the Dirac fields ψ, ψ^C have exactly the opposite charge to the corresponding particles.

We now illustrate why the existence of antiparticles with these properties is necessary to preserve causality [41, 42]. The Wightman propagator $S_+(x, y)$ describing the propagation of an electron from y to x is

---

[1] See Ref. [41] for our conventions and the various identities amongst the spinor quantities used here.

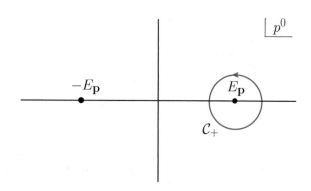

Fig. 2.1 The contour C_+ wraps around the pole at $p^0 = E_\mathbf{p}$ in the complex p^0 plane, defining the Wightman propagator $S_+(x, y)$

$$S_{+\alpha\beta}(x, y) = \langle 0|\psi_\alpha(x)\,\bar\psi_\beta(y)|0\rangle$$

$$= \int \frac{d^3\mathbf{p}}{(2\pi)^3}\frac{1}{\sqrt{2E_\mathbf{p}}} \int \frac{d^3\mathbf{q}}{(2\pi)^3}\frac{1}{\sqrt{2E_\mathbf{q}}} \sum_{s,s'} \langle 0|a_\mathbf{p}^s a_\mathbf{q}^{s'\dagger}|0\rangle\, u_\alpha^s(p)u_\beta^{s'}(q)\, e^{-ip.x+iq.y}$$

$$= \int \frac{d^3\mathbf{p}}{(2\pi)^3}\frac{1}{2E_\mathbf{p}}\,(\gamma.p + m)_{\alpha\beta}\, e^{-ip.(x-y)}, \tag{2.6}$$

with p^0 as defined previously. The last line follows using the anticommutation relations

$$\{a_\mathbf{p}^s, a_\mathbf{q}^{s'\dagger}\} = (2\pi)^3\delta^3(\mathbf{p} - \mathbf{q})\delta^{ss'}, \tag{2.7}$$

and the identity (2.4). This may also be written in a form which makes clear it is a solution of the homogeneous Dirac equation, viz.

$$S_+(x, y) = -i \int_{C_+} \frac{d^4 p}{(2\pi)^4} \frac{\gamma.p + m}{p^2 - m^2}\, e^{-ip.(x-y)}, \tag{2.8}$$

where the contour C_+ in the p^0 plane wraps completely around the pole at $p^0 = E_\mathbf{p}$ on the positive real axis, as shown in Fig. 2.1. The corresponding Green functions, including the Feynman propagator, are similarly defined with different contours around the poles.[2]

Similarly, we have

$$S_{-\alpha\beta}(x, y) = \langle 0|\bar\psi_\alpha(x)\,\psi_\beta(y)|0\rangle$$

$$= \int \frac{d^3\mathbf{p}}{(2\pi)^3}\frac{1}{2E_\mathbf{p}}\,(\gamma^T.p - m)_{\alpha\beta}\, e^{-ip.(x-y)}, \tag{2.9}$$

---

[2]Closed contours wrapping round either or both of the poles give solutions of the homogeneous Dirac equation (including the Wightman and anticommutator (Pauli–Jordan) functions), whereas open contours along the real p^0 axis diverting above or below the poles give Green functions, for example the retarded/advanced Green functions or the Feynman propagator.

with the contributions coming from the positron creation and annihilation operators $b_{\mathbf{p}}^s$, $b_{\mathbf{p}}^{s\dagger}$ and the spinors $v_\alpha^s(p)$. In the same way, the Wightman function for the charge conjugate fields is

$$
\begin{aligned}
S_{+\alpha\beta}^{\mathsf{C}}(x, y) &= \langle 0|\psi_\alpha^{\mathsf{C}}(x)\,\bar{\psi}_\beta^{\mathsf{C}}(y)|0\rangle \\
&= \int \frac{d^3\mathbf{p}}{(2\pi)^3}\frac{1}{2E_{\mathbf{p}}}\,(\gamma.p + m)_{\alpha\beta}\,e^{-ip.(x-y)},
\end{aligned}
\tag{2.10}
$$

derived from the positron operators and the $u_\alpha^s(p)$ spinors.

One way to phrase the requirement of causality in QFT (usually called 'microcausality' in this context) is to demand that the VEV of the anti-commutator of two spinor fields should vanish when they are spacelike separated. From the results above, still taking $p^0 = E_{\mathbf{p}}$, we find

$$
\langle 0|\{\psi_\alpha(x), \bar{\psi}_\beta(y)\}|0\rangle = (i\gamma.\partial_x + m)\int \frac{d^3\mathbf{p}}{(2\pi)^3}\frac{1}{2E_{\mathbf{p}}}\left(e^{-ip.(x-y)} - e^{ip.(x-y)}\right)
$$

$$
\to\ 0\ \text{ for }(x-y)^2 < 0.
\tag{2.11}
$$

To see this, note that for spacelike $(x - y)$ *only*, we can transform $(x - y) \to -(x - y)$ by a Lorentz transformation and rotation, then since the integration measure is Lorentz invariant it follows that the two terms cancel.[3]

The key point is that this cancellation requires the existence of antiparticles with the *exact* mass and spin, and opposite charge, of the corresponding particles. (Recall the second term arises from the positron operators $b_{\mathbf{p}}^s$, $b_{\mathbf{p}}^{s\dagger}$ whereas the first term involves the electron operators $a_{\mathbf{p}}^s$, $a_{\mathbf{p}}^{s\dagger}$.) Any differences, however small, of these properties would entail the violation of causality.

Notice, however, the key role of Lorentz invariance in this conclusion. We return to this in Sect. 2.3 where we discuss the possibility of Lorentz breaking.

2.2 Lorentz Symmetry and CPT

The preceding discussion has considered a QFT formulated on Minkowski spacetime, obeying the natural symmetries of Lorentz invariance $SO(1, 3)$ and spacetime translations T^4. This was sufficient to demonstrate that the existence of antiparticles with the exact mirror properties of their particles is required to preserve causality. Notice that as yet there has been no need to invoke CPT symmetry.

---

[3] We have illustrated this for free fields for simplicity. The argument could be generalised for interacting theories including self-energy contributions by the substitution $\gamma.p + m \to A(p^2)\gamma.p + B(p^2)$, etc.

Lorentz symmetry played the key role. Spin 1/2 fermions arise as spinor representations of the Lorentz group, or more precisely its double cover $SL(2, C)$. Left and right-handed Weyl spinors transform under the $(\frac{1}{2}, 0)$ and $(0, \frac{1}{2})$ representations of $SL(2, C)$, while the Dirac spinor describing the electron/positron is the representation $(\frac{1}{2}, 0) \oplus (0, \frac{1}{2})$. The fact that spin 1/2 particles are described as Lorentz group representations will be crucial when when we introduce gravity, especially in reference to CPT.

In addition to the continuous symmetries, we may also discuss the discrete transformations of charge conjugation (C), parity (P) and time reversal (T) in the Dirac theory. While these are independently symmetries of QED, in the full standard model each is known to be broken. However, the combination CPT has a special status and is conserved in any local, relativistic QFT.

The actions of C, P and T on the creation and annihilation operators are simple:

$$a_{\mathbf{p}}^s \xrightarrow{C} b_{\mathbf{p}}^s, \qquad a_{\mathbf{p}}^s \xrightarrow{P} a_{-\mathbf{p}}^s, \qquad a_{\mathbf{p}}^s \xrightarrow{T} a_{-\mathbf{p}}^{-s},$$
$$b_{\mathbf{p}}^s \xrightarrow{C} a_{\mathbf{p}}^s, \qquad b_{\mathbf{p}}^s \xrightarrow{P} -b_{-\mathbf{p}}^s, \qquad b_{\mathbf{p}}^s \xrightarrow{T} b_{-\mathbf{p}}^{-s}.$$

$$(2.12)$$

Inserting these relations into the expansion (2.2) for the fields, and using appropriate identities for the $u_\alpha^s(p)$ and $v_\alpha^s(p)$ spinors [41], we find the corresponding, but less transparent, relations (just writing the combined PT transformation for simplicity):

$$\psi(x) \xrightarrow{C} \psi^C(x) = C\bar{\psi}^T(x),$$
$$\psi(x) \xrightarrow{PT} \psi^{PT}(x) = P\psi(-x),$$

$$(2.13)$$

where C is as above and $P = -\gamma^0\gamma^1\gamma^3$, with similar expressions for $\bar{\psi}$.[4] Combining these, we find the key CPT transformations:

$$\psi^{CPT}(x) = P\psi^C(-x) = -\gamma^5\psi^*(-x),$$
$$\bar{\psi}^{CPT}(x) = -\bar{\psi}^C(-x)P = \bar{\psi}^*(-x)\gamma^5.$$

$$(2.14)$$

Writing ψ^{CPT} explicitly in terms of the creation and annihilation operators,

$$\psi_\alpha^{CPT}(x) = \int \frac{d^3\mathbf{p}}{(2\pi)^3} \frac{1}{\sqrt{2E_\mathbf{p}}} \sum_s \left(-b_\mathbf{p}^{-s} u_\alpha^s(p)^* e^{ip.x} + a_\mathbf{p}^{-s\dagger} v_\alpha^s(p)^* e^{-ip.x}\right). \quad (2.15)$$

---

[4]In general, there are phases associated with each of the transformations C, P and T, whose product is constrained to be 1. These phases are all set to 1 here, but we need to retain the relative minus sign in the parity transformations of fermions and antifermions in (2.12) to ensure the correct parity assignments for fermion-antifermion bound states [41].

Note that C and P are realised by linear, unitary operators whereas T, and therefore CPT, is an anti-linear, anti-unitary transformation.

We can now prove a fundamental result using CPT invariance to relate the physics of particles and antiparticles. If CPT is a symmetry of the QFT, and not spontaneously broken by the vacuum state, then the Wightman propagator satisfies

$$\langle 0 | \psi(x)\, \bar{\psi}(y) | 0 \rangle \;=\; \langle 0 | \psi^{\mathsf{CPT}}(x)\, \bar{\psi}^{\mathsf{CPT}}(y) | 0 \rangle^* \,. \tag{2.16}$$

The complex conjugate on the r.h.s. arises because of the anti-unitary property of CPT. Evaluating this, we find[5]

$$
\begin{aligned}
\langle 0 | \psi^{\mathsf{CPT}}(x)\, \bar{\psi}^{\mathsf{CPT}}(y) | 0 \rangle^* &= \left(-\boldsymbol{P}\, \langle 0 | \psi^{\mathsf{C}}(-x)\, \bar{\psi}^{\mathsf{C}}(-y) | 0 \rangle\, \boldsymbol{P} \right)^* \\[4pt]
&= \left(-\boldsymbol{P} \int \frac{d^3\mathbf{p}}{(2\pi)^3}\, \frac{1}{2E_\mathbf{p}}\, \left(A(p^2)\gamma \cdot p + B(p^2) \right)\, \boldsymbol{P}\, e^{ip\cdot(x-y)} \right)^* \\[4pt]
&= \left(\int \frac{d^3\mathbf{p}}{(2\pi)^3}\, \frac{1}{2E_\mathbf{p}}\, \left(A(p^2)\gamma^* \cdot p + B(p^2) \right)\, e^{ip\cdot(x-y)} \right)^* \\[4pt]
&= \langle 0 | \psi^{\mathsf{C}}(x)\, \bar{\psi}^{\mathsf{C}}(y) | 0 \rangle \,. \tag{2.17}
\end{aligned}
$$

We have allowed here for self-energies through $A(p^2)$, $B(p^2)$ so this result holds for an interacting theory, not simply for free fields. Crucially, the derivation assumes both Lorentz invariance *and* translation invariance (since we have assumed the propagator is a function of $(x-y)$ in the second step). Under these assumptions, which must be reconsidered in the presence of gravity [43], we have therefore shown that CPT invariance implies that the propagators of particles and antiparticles in Minkowski spacetime are identical, i.e.

$$\langle 0 | \psi(x)\, \bar{\psi}(y) | 0 \rangle \;=\; \langle 0 | \psi^{\mathsf{C}}(x)\, \bar{\psi}^{\mathsf{C}}(y) | 0 \rangle \,. \tag{2.18}$$

The general relation between Lorentz invariance and CPT symmetry is expressed in the famous CPT theorem [3–6]. The theorem may be proved rigorously in axiomatic field theory, but is fundamentally simple. The essential statement is that in a *local, Lorentz invariant* QFT, CPT is an exact symmetry. A simple proof is given in [42], where it is shown, based on the transformations (2.14) together with their equivalents for scalar and vector fields, that any local product of scalar, spinor and vector fields is necessarily invariant under the combined CPT transformation. This simply means that any Lorentz invariant interaction we may write down in an effective Lagrangian is CPT invariant.

To illustrate this, we can construct a table of the basic fermion bilinear operators relevant to the analysis of low-energy antimatter experiments, together with their C, P, T and CPT transformations [41]. It is clear from the table that the Lorentz scalar operators are CPT even, in accord with the CPT theorem.

---

[5]Notice that the spin flip under CPT in the creation and annihilation operators is not apparent in the propagators themselves because of the spin sum.

2.3 Breaking Lorentz Invariance and CPT

So far we have reviewed how Lorentz invariance, CPT and translation invariance are fundamental to the local, relativistic QFTs which successfully describe particle physics in Minkowski spacetime. The fact that quantum fields are representations of the Lorentz group $SO(1, 3)$ (or for spinor fields, $SL(2, C)$) is essential in the first place to have particle states with definite masses and spins, since these are the eigenvalues of the two Casimir operators of the Lorentz group. The existence of antiparticles with exactly the same mass and spin and opposite charge was then shown to be required to maintain causality. We also presented a direct proof that CPT symmetry, together with Lorentz and translation invariance, implies the equality of the propagators for particles and antiparticles. Finally, the CPT theorem ensures that CPT is an exact symmetry of any local, Lorentz invariant QFT.

The question then arises how this tightly-woven theoretical structure could be unravelled in the event that experiments contradicted its clear predictions, e.g. in violating charge neutrality in antihydrogen or finding a difference in the atomic spectroscopy of hydrogen and antihydrogen?

The simplest way to describe a possible breaking of Lorentz invariance, while preserving the fundamental structure of causal fields in representations of the Lorentz group, is to write a phenomenological Lagrangian including operators which are not Lorentz invariant. For QED, we therefore consider

$$\mathcal{L}_{\text{LV}} = \mathcal{L}_{\text{QED}} - \tfrac{1}{4}(k_F)_{\mu\rho\nu\sigma}F^{\mu\rho}F^{\nu\sigma} + \tfrac{1}{2}(k_{AF})^{\rho}\epsilon_{\rho\mu\nu\sigma}A^{\mu}F^{\nu\sigma}$$
$$- a_{\mu}\bar{\psi}\gamma^{\mu}\psi - b_{\mu}\bar{\psi}\gamma^5\gamma^{\mu}\psi - \tfrac{1}{2}H_{\mu\nu}\bar{\psi}\sigma^{\mu\nu}\psi + c^{\mu\nu}i\bar{\psi}\gamma_{\mu}D_{\nu}\psi + d^{\mu\nu}i\bar{\psi}\gamma_5\gamma_{\mu}D_{\nu}\psi \,.$$
$$(2.19)$$

This is just the restriction to QED[6] of the full "Standard Model Extension" (SME) of Kostelecký and collaborators [39, 40]. This has been the subject of an extensive programme of research over many years, with stringent experimental bounds being established on many of the Lorentz-violating couplings [44]. Note the profligacy of the parameter count in these theories once Lorentz symmetry is lost—the QED Lagrangian $\mathcal{L}_{\text{LV}}$ alone has some 70 independent parameters.

The experimental consequences of some of these additional Lorentz-violating interactions for antimatter will be discussed below. First, we need some comments on the theoretical basis of (2.19).

A key point is that only *local* operators have been included. Locality is one of the axioms underlying the CPT theorem, so if locality were not valid this would break the link between Lorentz invariance and CPT. String theory may initially be thought to exploit this loophole, but even here the theory at low energies, well below the string or Planck scales, is still described by a local effective Lagrangian

---

[6]Note that here we omit three further CPT odd operators, $ie^{\mu}\bar{\psi}D_{\mu}\psi$, $f^{\mu}\bar{\psi}\gamma^5 D_{\mu}\psi$ and $\tfrac{i}{2}g^{\lambda\mu\nu}\bar{\psi}\sigma_{\mu\nu}D_{\lambda}\psi$, which could arise in QED alone but which are not obtained as a restriction of the SME [40].

satisfying the CPT theorem. A form of non-locality is intrinsic to quantum mechanics through entanglement, but again this is not of a form that impacts on the CPT theorem. Other more exotic mechanisms for CPT violation, *e.g.* involving non-trivial spacetime topology [45, 46], are also not relevant here. Note, however, the ingenious proposal of [47, 48] (for a review, see [49]) of a non-local, CPT violating but Lorentz invariant extension of QED. At energies well below the non-locality scale, this theory exhibits different effective masses for the electron and positron, with the obvious consequences for the hydrogen and antihydrogen spectra. The theory is non-unitary, while the realisation of causality and microcausality is subtle due to the non-locality, so it is best regarded as itself only an effective theory [48, 50] from the point of view of analysing antimatter experiments at the energy scales of atomic physics. We will not pursue such non-local theories further in this book. Instead, we simply accept locality and the CPT theorem, in particular that a violation of CPT necessarily entails a violation of Lorentz invariance.

Next, note that it is implicit in (2.19), where the fields $\psi(x)$ are the usual spinor representations of the Lorentz group and admit the decomposition (2.2) in terms of creation and annihilation operators, that the charge and mass of particles and their antiparticles are identical. This is a key feature of the SME and allows it, with the provisos below, to have an interpretation as a causal theory. We could imagine instead writing the expansion (2.2) with different masses for the particles and antiparticles, so the momenta associated with the operators $a_{\mathbf{p}}^s$ and $b_{\mathbf{p}}^s$ satisfied energy-momentum relations with a different mass. This was investigated by Greenberg in [51], with the conclusion, unsurprising in view of the discussion of causality in Sect. 2.1 and in what follows below, that such a 'theory' would not be causal and in a certain sense non-local. This justifies the restriction to the less destructive breaking of Lorentz and CPT invariance parametrised in the SME Lagrangian (2.19).

Finally, we should distinguish between *spontaneous* and *explicit* breaking of Lorentz invariance. One way to view the coefficients a_μ, b_μ, ... in (2.19) is as the spontaneous Lorentz-violating VEVs of vector or tensor fields in some fundamental Lorentz invariant theory. $\mathcal{L}_{\mathrm{LV}}$ would then represent an effective Lagrangian describing this theory at low energies where the dynamics of these new fields can be neglected.

More generally, the philosophy of low-energy effective Lagrangians is to write an expansion in terms of operators of increasing dimension. Operators of dimension >4 will have coefficients with dimensions of inverse powers of mass and will therefore be suppressed by the mass scale characteristic of some unknown dynamics at high energy. (A familiar example is the chiral Lagrangian for low-energy QCD.) This motivates us to restrict $\mathcal{L}_{\mathrm{LV}}$ only to the soft or renormalisable Lorentz-violating operators, with dimension ≤ 4, shown in (2.19). With this restriction, the theory is known as the *minimal* SME. If operators of dimension > 4 are included, it is referred to as non-minimal. We consider some examples of non-minimal operators in Sect. 3.2.2.

Alternatively, we can simply break Lorentz invariance *explicitly*, in which case the coefficients a_μ, b_μ, ... are merely constants with no relation to underlying covariant fields. For low-energy phenomenology the distinction is not so important, but the

Table 2.1 C, P, T and CPT transformations for the basic fermion bilinear operators. The notation is: $(-1)^\mu = 1$ for $\mu = 0$ and -1 for $\mu = 1, 2, 3$

	$\bar\psi\psi$	$i\bar\psi\gamma^5\psi$	$\bar\psi\gamma^\mu\psi$	$\bar\psi\gamma^\mu\gamma^5\psi$	$\bar\psi\sigma^{\mu\nu}\psi$
C	$+1$	$+1$	-1	$+1$	-1
P	$+1$	-1	$(-1)^\mu$	$-(-1)^\mu$	$(-1)^\mu(-1)^\nu$
T	$+1$	-1	$(-1)^\mu$	$(-1)^\mu$	$-(-1)^\mu(-1)^\nu$
CPT	$+1$	$+1$	-1	-1	$+1$

issue of whether (2.19) is to be viewed as a complete theory in itself or simply as an effective Lagrangian with a Lorentz-invariant UV completion, is important for its theoretical interpretation, particularly as regards causality.

As can be seen from Table 2.1, and in accord with the CPT theorem, the Lorentz-violating operators in (2.19) divide into those that conserve CPT and those that violate it, while of course all the operators in the original Lorentz-invariant $\mathcal{L}_{\text{QED}}$ are CPT conserving. Of the complete set of Lorentz-violating operators, those with couplings a_μ, b_μ, k_{AF} are CPT odd, while those with $c^{\mu\nu}, d^{\mu\nu}, H^{\mu\nu}, k_F$ are CPT even. Also note that a_μ couples to a C odd operator, while the corresponding operator with b_μ is C even.

We can analyse $\mathcal{L}_{\text{LV}}$ in the same way as presented above for the standard Lorentz-invariant Dirac theory. To illustrate some of the theoretical issues, we show some results in the especially simple case where only a_μ is taken to be non-zero.

The modified Dirac equation is then

$$(i\gamma.\partial - \gamma.a - m)\,\psi = 0, \tag{2.20}$$

while the expansion of the field $\psi(x)$ in creation and annihilation operators becomes[7]

$$\psi(x) = \int \frac{d^3\mathbf{p}}{(2\pi)^3} \sum_s \left(\frac{1}{\sqrt{2E_{\mathbf{p-a}}}}\, a^s_{\mathbf{p-a}}\, u^s \left(E_{\mathbf{p-a}}, \mathbf{p-a}\right) e^{-ip_u.x} \right.$$
$$\left. + \frac{1}{\sqrt{2E_{\mathbf{p+a}}}}\, b^{s\dagger}_{\mathbf{p+a}}\, v^s \left(E_{\mathbf{p+a}}, \mathbf{p+a}\right) e^{ip_v.x} \right), \tag{2.21}$$

where $p_u = \left(E_{\mathbf{p-a}} + a^0, \mathbf{p}\right)$ and $p_v = \left(E_{\mathbf{p+a}} - a^0, \mathbf{p}\right)$, with $E_{\mathbf{p-a}} = \sqrt{(\mathbf{p-a})^2 + m^2}$. The dispersion relations are:

$$(p_u - a)^2 - m^2 = 0, \qquad (p_v + a)^2 - m^2 = 0, \tag{2.22}$$

while the spinors u, v are the same functions as usual, with shifted arguments.

---

[7]Note that in this special case, we could equally well move the $\mathbf{a}$ dependence from the spinors entirely into the exponent through a change of integration variable $\mathbf{p} \to \mathbf{p} - \mathbf{a}$, though this will not be possible in the general Lorentz-violating theory.

Fig. 2.2 The contours C_+^u and C_-^u in the complex p^0 plane relevant for calculating the Wightman and anticommutator functions in the Lorentz-violating theory with a_μ non-zero. The abbreviated notation for the poles is $\tilde{E}_\mathbf{p}^{-+} = E_{\mathbf{p}-\mathbf{a}} + a^0$ and $\tilde{E}_\mathbf{p}^{--} = E_{\mathbf{p}-\mathbf{a}} - a^0$, etc.

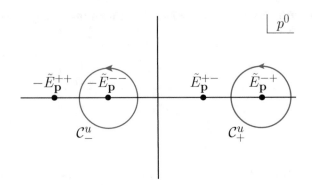

Analogous expressions can be written for the C and CPT conjugates $\psi^C(x) = C\bar{\psi}^T(x)$ and $\psi^{CPT}(x) = -\gamma^5\psi^*(-x)$, making the transformations (2.12) for the creation/annihilation operators together with the substitution $a_\mu \to -a_\mu$ since this is C odd (see Table 2.1).

The Wightman propagators are easily found:

$$S_+(x,y) = \langle 0|\psi(x)\,\bar{\psi}(y)|0\rangle$$

$$= \int \frac{d^3\mathbf{p}}{(2\pi)^3} \frac{1}{2E_{\mathbf{p}-\mathbf{a}}} \left(\gamma.\left(E_{\mathbf{p}-\mathbf{a}}, \mathbf{p}-\mathbf{a}\right) + m\right) e^{-ip_u.(x-y)}$$

$$= -i \int_{C_+^u} \frac{d^4p}{(2\pi)^4} \frac{\gamma.(p-a)+m}{(p-a)^2-m^2} e^{-ip.(x-y)}, \tag{2.23}$$

where C_+^u circles the pole at $p^0 = E_{\mathbf{p}-\mathbf{a}} + a^0$ (Fig. 2.2).

The equivalent result for $\langle 0|\psi^C(x)\,\bar{\psi}^C(y)|0\rangle$ follows with the substitution $a_\mu \to -a_\mu$, the poles being shifted accordingly. Unsurprisingly, in this Lorentz-violating theory the Wightman propagators for the fields ψ and ψ^C are different—electrons and positrons have different dispersion relations, given in (2.22), and propagate differently.

The derivation in (2.17) for the Wightman propagator for ψ^{CPT} remains unchanged, so we have

$$\langle 0|\psi^C(x)\,\bar{\psi}^C(y)|0\rangle = \langle 0|\psi^{CPT}(x)\,\bar{\psi}^{CPT}(y)|0\rangle^*$$

$$\neq \langle 0|\psi(x)\,\bar{\psi}(y)|0\rangle, \tag{2.24}$$

as expected in a theory with CPT violation.

This immediately raises the issue of causality. Given the rôle of Lorentz invariance and the exact equivalence of the properties of particles and antiparticles in establishing microcausality in Sect. 2.1, we have to question whether the introduction of the Lorentz-violating terms in $\mathcal{L}_{LV}$ necessarily violates microcausality.

For the simple case with only a_μ non-zero, this is readily answered. After some calculation we can explicitly write the anticommutator function as

$$\langle 0|\{\psi(x), \bar{\psi}(y)\}|0\rangle = (i\gamma.\partial_x - \gamma.a + m) \int \frac{d^3\mathbf{p}}{(2\pi)^3} \left(\frac{1}{2E_{\mathbf{p-a}}} e^{-ip_u.(x-y)} - \frac{1}{2E_{\mathbf{p+a}}} e^{ip_v.(x-y)} \right),$$
(2.25)

which follows from integrating around both the contours C_+^u and C_-^u in Fig. 2.2. It is critical here that these contours circle the poles which are both shifted to the right by $+a_0$. After some changes of variable, it then follows that

$$\langle 0|\{\psi(x), \bar{\psi}(y)\}|0\rangle = e^{-ia.(x-y)} (i\gamma.\partial_x + m) \int \frac{d^3\mathbf{p}}{(2\pi)^3} \frac{1}{2E_{\mathbf{p}}} \left(e^{-ip.(x-y)} - e^{ip.(x-y)} \right) \Big|_{p^0 = E_{\mathbf{p}}},$$
(2.26)

where, as a result of the specification of the contours above, the a_μ dependence factorises. This is the key step. The remaining integral is then exactly as appeared in the Lorentz-invariant theory, and vanishes for spacelike $(x - y)$.

Perhaps counter-intuitively, we therefore find that microcausality continues to hold even in the theory with Lorentz and CPT violation induced by a non-vanishing a_μ coupling, despite the apparently different propagation characteristics of electrons and positrons. In fact, there is a reason for this [39]. The couplings a^μ in $\mathcal{L}_{LV}$ can be removed by a field redefinition if we write $\psi(x) = \exp(-ia.x)\xi(x)$ and rewrite the Lagrangian in terms of the new field $\xi(x)$. Physical electrons and positrons are then defined in terms of creation and annihilation operators in $\xi(x)$.

In general, the dispersion relations following from $\mathcal{L}_{LV}$ are *quartic* in the momenta, which doubles the number of poles in the complex p^0 plane in the calculation of the Wightman propagators and other Green functions. Nevertheless, a similar argument to that shown above for the simpler case of a_μ demonstrates that the Lorentz-violating theory with the couplings b_μ or $H_{\mu\nu}$ also satisfies microcausality [39, 52].

The situation is more subtle when the couplings $c^{\mu\nu}$, $d^{\mu\nu}$ to operators with derivatives are present. In this case, the poles may not lie on the real p^0 axis, and it is easy to identify special cases (e.g. $c^{00} < 0$) where microcausality is violated and superluminal phase velocities arise for electrons/positrons with momentum exceeding some threshold [39, 52]. This is also evident for certain of the photon couplings $(k_F)_{\rho\sigma\mu\nu}$ and $(k_{AF})^\rho$. This means that the theory (2.19) as it stands is *not* causal. However, if $\mathcal{L}_{LV}$ is regarded only as a low-energy effective Lagrangian, then it remains possible that it may admit a causal UV completion. As emphasised in [53, 54], and the series of papers [55–58] in a gravitational context, it is the UV limit of a QFT which determines whether or not it is causal. In particular, in a theory with non-trivial dispersion relations, causality requires the high-momentum limit of the phase velocity to be less than c (*not* the group velocity, which is irrelevant for causality in general). There exist perfectly causal QFTs which nevertheless have a low-energy effective Lagrangian exhibiting superluminal propagation and apparent microcausality violation.

It follows that, except for the restricted case with only a_μ, b_μ, $H_{\mu\nu}$ non-zero, causality requires that $\mathcal{L}_{LV}$ is to be regarded as an effective Lagrangian only. Imposing causality of the fundamental theory of nature does not then preclude Lorentz and CPT violations from arising in low-energy experiments. Indeed, an experimental measurement indicating a non-vanishing value for one of the Lorentz-violating couplings in $\mathcal{L}_{LV}$ may provide an important clue as to the nature of the fundamental theory at high, perhaps Planckian, energies.

2.4 General Relativity and Equivalence Principles

So far, we have not considered gravity. However, the precision now being attained by laboratory experiments involving antimatter means that a full understanding of their implications for fundamental physics requires us to consider gravitational effects.

The established, and phenomenally successful, theory of gravity in classical physics is general relativity (GR). It follows that we should use GR from the outset as the framework to discuss gravitational effects in antimatter experiments, not resort immediately to ad-hoc descriptions based on the superseded concepts of Newtonian theory, e.g. that gravity couples to mass rather than, as in GR, the energy-momentum tensor.

The essential insight of GR is to replace the notion of an independent gravitational force by the introduction of curved spacetime as the arena on which the laws of nature are formulated. Here, spacetime is taken to be a pseudo-Riemannian manifold,[8] i.e. it admits a metric of the standard quadratic form,

$$ds^2 = g_{\mu\nu}dx^\mu dx^\nu \ . \tag{2.27}$$

The fundamental theorem of Riemannian geometry then asserts that there is a unique, torsion-free, metric-preserving connection (the Levi-Civita connection) enabling parallel transport on the manifold. This connection has the property that around any given point on the spacetime manifold we can choose a local orthonormal frame (Riemann normal coordinates) in which *locally* the metric tensor reduces to the Minkowski metric and the Christoffel symbols (which depend on the first derivatives of the metric) vanish. In this sense, we can characterise the essential feature of the spacetime of GR as "locally flat". The key symmetry principle, therefore, is that while in the presence of gravity we must abandon global Lorentz invariance and translation invariance, we maintain *local* Lorentz invariance.

In physics terms, this implements geometrically the requirement that at each point in spacetime, there exists a *local inertial frame* (LIF). "Local" in this context means relative to the scale at which tidal gravitational effects depending on the spacetime curvature (which depends on the second derivatives of the metric) become important. The existence of LIFs is the central principle on which GR is based, and dictates the choice of Riemannian manifolds as the geometric description of spacetime.

---

[8]The terminology "pseudo" simply indicates the Minkowski signature of the metric $(-+++)$ rather than the Euclidean signature. In what follows, we just refer to the Minkowski signature manifolds as Riemannian for simplicity.

To couple matter to gravity we therefore have to formulate QFT on a curved spacetime. We outline this here for a Dirac field. The immediate problem, as discussed earlier, is that the spinor field is a representation of the $SL(2, C)$ covering group of the Lorentz group. But Lorentz invariance is no longer a global symmetry of spacetime in GR as it is in special relativity. However, because of the local Lorentz symmetry, we can still describe fermions by defining spinor fields with respect to the local orthonormal frame at each point in the curved spacetime. This is done using the vierbein formalism. We define the vierbein $e_a{}^\mu(x)$ such that

$$\eta_{ab} = g_{\mu\nu}(x)e_a{}^\mu(x)e_b{}^\nu(x), \tag{2.28}$$

where $\eta_{ab} = \text{diag}(-1, 1, 1, 1)$ is the Minkowski metric. The coupled gravity–Dirac action in GR then becomes (ignoring the cosmological constant, which will not play a rôle here)

$$S = \int d^4x \sqrt{-g} \left(\frac{R}{16\pi G} + \bar{\psi} \left(i\gamma^a e_a{}^\mu D_\mu - m \right) \psi \right). \tag{2.29}$$

The covariant derivative on spinors is

$$D_\mu \psi = \left(\partial_\mu - \tfrac{i}{4}\omega_\mu{}^{bc}\sigma_{bc} \right) \psi, \tag{2.30}$$

where $\sigma^{ab} = \tfrac{i}{2}\left[\gamma^a, \gamma^b\right]$ and the spin connection is defined as $\omega_\mu{}^b{}_c = e_\nu{}^b \left(\partial_\mu e_c{}^\nu + \Gamma^\nu_{\mu\rho}e_c{}^\rho\right)$.

The first key point here is that it is only the existence of local orthonormal frames, itself a property of Riemannian spacetime, that allows us to describe spin 1/2 particles and Dirac fields in the usual way at all.

Next, note that in this conventional GR action, the gravity-matter coupling is through the connection *only*, not the curvature. Since the Christoffel symbols $\Gamma^\nu_{\mu\rho} \sim 0$ in a *local* inertial frame (i.e. in Riemann normal coordinates), it follows that in a LIF $D_\mu \psi \sim \partial_\mu \psi$ and the Dirac action takes its special relativistic form. This is how conventional GR implies the Strong Equivalence Principle, viz. that the laws of physics take their special relativistic form in the local inertial frame at any given point in spacetime.

We can also note immediately the implications for CPT symmetry. Since in GR we only have local Lorentz invariance, it follows that the discrete symmetries P, T and therefore CPT are only defined as transformations in the local Minkowski space at each spacetime point. (See e.g. Ref. [59] for further elaboration.) In particular, they have nothing to do with the extended nature of the curved spacetime. P or T invariance of the action (2.29) does not involve any sort of space or time reflection symmetry of the curved spacetime manifold.

The dynamical Einstein equations are derived by varying the action (2.29) with respect to the metric $g_{\mu\nu}$, giving

$$G_{\mu\nu} \equiv R_{\mu\nu} - \tfrac{1}{2} R g_{\mu\nu} = 8\pi G T_{\mu\nu}, \tag{2.31}$$

where $T_{\mu\nu} = \tfrac{i}{2}\bar{\psi}\gamma^a e_a{}^{\{\mu} D^{\nu\}}\psi$ is the covariantly-conserved energy-momentum tensor for the Dirac field and $R_{\mu\nu}(R)$ is the Ricci curvature tensor (scalar). Note that the Einstein tensor $G_{\mu\nu}$ is automatically conserved by virtue of the Bianchi identity, matching the r.h.s. of (2.31).

This illustrates another key point. In GR, the gravitational field couples to the energy-momentum tensor, *not* the mass. It is described by a tensor field, the metric $g_{\mu\nu}(x)$, which in quantum theory corresponds to a massless spin 2 particle (the graviton) propagating with the speed of light, as has been beautifully confirmed by the recent discovery of gravitational waves [60]. The interaction is always attractive.

This is in marked contrast to the old Newtonian view, based on the force equation $F = G m_1 m_2 / r^2$. Apart from invoking action at a distance, this encourages the misleading idea of mass as a 'gravitational charge' by comparing with the analogous equation in electrostatics. However, in electromagnetism the force is mediated by a spin 1 particle, the photon, and so the force between charges may be either attractive or repulsive. This is not the case with gravity. Moreover, the frequently invoked distinction between 'gravitational' and 'inertial' mass is not relevant once we have abandoned the obsolete Newtonian force equation and adopted GR as the theory of gravity. There is only one mass in the GR action (2.29), viz. the Lorentz invariant particle mass in the LIF in which the spinor field is defined.

We now collect some elementary consequences of GR that we will need later in analysing the experiments. First, it follows from the conservation of the energy-momentum tensor that free particles follow geodesics in curved spacetime. That is,

$$\frac{d^2 x^{\mu}}{d\lambda^2} + \Gamma^{\mu}_{\rho\sigma} \frac{dx^{\rho}}{d\lambda} \frac{dx^{\sigma}}{d\lambda} = 0, \tag{2.32}$$

where $x^{\mu}(\lambda)$ describes the particle trajectory parametrised by the affine parameter λ. Clearly, in the LIF where $\Gamma^{\mu}_{\rho\sigma} \sim 0$, this path becomes a straight line.

The geodesic equation is simply derived from the covariant conservation of the energy-momentum tensor, $\nabla_{\mu} T^{\mu\nu} = 0$, using the explicit form of the energy-momentum tensor for a free particle (see e.g. Ref. [61]). An alternative derivation, which we shall use later in discussing possible WEPff violations, is to find the equation of motion by extremising the action for a point particle in curved spacetime, viz.

$$S = -m \int ds = -m \int d\lambda \sqrt{g_{\mu\nu} \frac{dx^{\mu}}{d\lambda} \frac{dx^{\nu}}{d\lambda}}. \tag{2.33}$$

A short calculation then shows,

$$\delta S = m \int d\lambda \left(\frac{ds}{d\lambda}\right)^{-1} \delta x_{\mu} \left[\frac{d^2 x^{\mu}}{d\lambda^2} + \frac{1}{2} g^{\mu\alpha} \left(g_{\alpha\rho,\sigma} + g_{\sigma\alpha,\rho} - g_{\rho\sigma,\alpha}\right) \frac{dx^{\rho}}{d\lambda} \frac{dx^{\sigma}}{d\lambda} \right]. \tag{2.34}$$

Identifying the second term as the Christoffel symbol, the geodesic equation follows as $\delta S/\delta x^\mu = 0$.

Note that the mass m occurs here simply as an overall factor in S and does not appear in the geodesic equation. Put otherwise, the *same* mass parameter multiplies both the "acceleration" term and the "gravity" term in the equation of motion. As we now see, this is how the essentially Newtonian formulation of the equivalence principle as the identity $m_i = m_g$ of the "inertial" and "gravitational" masses is realised.

To show this, we reproduce the equation of motion for a particle falling in the Earth's gravitational field. The gravitational field in the exterior region of a spherically symmetric mass M is described by the Schwarzschild metric,

$$ds^2 = -\left(1 - \frac{2GM}{r}\right)dt^2 + \left(1 - \frac{2GM}{r}\right)^{-1} dr^2 + r^2 d\theta^2 + r^2 \sin^2(\theta)d\phi^2 .$$

(2.35)

In the weak field limit, we may write the metric in the form $g_{\mu\nu} = \eta_{\mu\nu} + h_{\mu\nu}$ and work to first order in $h_{\mu\nu}$. Now, with the standard simplifications for a slow-moving (non-relativistic) particle, and since $\Gamma^\mu_{00} = -\frac{1}{2}g^{\mu j}g_{00,j}$ for a static metric, the geodesic equation (2.32) quickly reduces to[9]

$$\frac{d^2 x^r}{dt^2} = \frac{1}{2}\partial_r h_{00},$$

(2.36)

for a radially falling particle. This matches the Newtonian equation with potential $U(r) = -\frac{1}{2}h_{00} = -GM/r$. (Note we are using $c = 1$ units throughout.) At the Earth's surface, $U = -GM/R \simeq -7 \times 10^{-10}$. Again, notice that the mass of the falling particle (being equal on both sides of (2.36)) cancels out from the equation of motion in accordance with the foundations of GR. This experimental prediction of GR therefore realises the universality of free-fall (WEPff).

An important issue in GR is to relate the coordinates used to describe the spacetime metric (and in which calculations are most readily performed) with the physical measurements of space and time made by individual observers. (See, e.g. [61] for a particularly clear account.) This gives rise to *gravitational time dilation* and will be a key factor in interpreting the results of spectroscopic frequency measurements in curved spacetime.

In theoretical terms, we regard each observer (whether freely-falling or not) as equipped with a local orthonormal frame $\hat{e}_a$ ($a = 0, 1, 2, 3$) whose components with respect to the coordinate basis define a vierbein $\hat{e}_a{}^\mu$ as described above. The timelike frame vector is chosen to lie along the wordline of the observer: specifically, $\hat{e}_0{}^\mu = u^\mu$ where $u^\mu = dx^\mu/d\sigma$ is the observer's 4-velocity and σ is the proper time along the observer's worldline (so $d\sigma^2 = -g_{\mu\nu}dx^\mu dx^\nu$ and the 4-velocity is normalised

---

[9]Note that Eq. (2.36) refers to the space and time coordinates as they appear in the Schwarzschild metric—as explained below, these have to be translated into the physical measurements made by an individual observer in order to confront with experiment. However, this translation only applies a correction of $O(GM/r)$ and may be neglected since the r.h.s. is already of that order.

by $g_{\mu\nu}u^\mu u^\nu = -1$). A spacetime coordinate interval dx^μ is then measured by this observer as the *projection onto this local orthonormal frame*. That is,

$$d\hat{x}^a = \eta^{ab}\hat{e}_b{}^\mu g_{\mu\nu}dx^\nu \equiv \hat{e}^a{}_\mu dx^\mu, \tag{2.37}$$

where $\hat{e}^a{}_\mu$ is the inverse vierbein field.

It follows immediately that if a frequency source, e.g. an atomic spectral transition, is moving through spacetime (so the coordinate interval to be measured is dx^μ), then a *comoving* observer will measure the corresponding time interval as $d\hat{t} = d\sigma$, the *proper time along the worldline* of the moving source. Explicitly, this fundamental observation follows from the formalism above as

$$d\hat{t} = \eta^{00}\,\hat{e}_0{}^\mu\,g_{\mu\nu}\,dx^\nu = -u^\mu\,g_{\mu\nu}\,dx^\nu = d\sigma\,. \tag{2.38}$$

Now consider the measurement of the time interval between two events at the same point P, separated by a coordinate interval dt. In the rest frame, the time measurement is simply the corresponding proper time interval $d\tau = \sqrt{-g_{00}}\,dt$, defined from the metric as $ds^2 = -d\tau^2$. In a moving frame, with coordinate velocity $v^i = dx^i/dt$, the measured time between the two events is, for a diagonal metric,

$$d\hat{t} = \eta^{00}\hat{e}_0{}^0\,g_{00}(P)\,dt = -u^0\,g_{00}(P)\,dt \equiv \frac{d\tau}{d\sigma}\,d\tau\,. \tag{2.39}$$

To evaluate this time dilation factor, note that

$$u^0 = \frac{dt}{d\sigma} = \left(-g_{00} - g_{ij}v^i v^j\right)^{-1/2}, \tag{2.40}$$

and so,

$$\frac{d\tau}{d\sigma} = \left(1 + g_{ij}v^i v^j/g_{00}\right)^{-1/2}. \tag{2.41}$$

We recognise that *locally* this factor is just the expression in curved spacetime coordinates of the special relativistic time dilation factor $\gamma(v^2)$ in the orthonormal frame at P, as follows from the vierbein definition $d\hat{x}^a = \hat{e}^a{}_\mu dx^\mu$ above.[10]

---

[10]As a further demonstration of consistency, we should find the same time dilation factor for the relativistically equivalent situation where we instead consider two spacetime events *on the worldline of the moving observer*, separated by coordinate interval dt and therefore $dx^i = v^i dt$. In the comoving frame, the measured time is the proper time interval along the worldline, $d\sigma$. In the stationary frame, the formalism above gives the measured time interval as

$$d\tilde{t} = -\tilde{u}^\mu\,g_{\mu\nu}(P)\,dx^\nu,$$

where $\tilde{u}^\mu$ is the 4-velocity of the stationary observer, that is $\tilde{u}^0 = dt/d\tau$ and $\tilde{u}^i = 0$. Since the metric is assumed diagonal, this reduces to

$$d\tilde{t} = -\tilde{u}^0\,g_{00}(P)\,dt = d\tau = \frac{d\tau}{d\sigma}\,d\sigma\,.$$

To illustrate this further, consider a frequency measurement at a given space position $r = r_O$ in the Schwarzschild metric. An observer (O) stationary at $r = r_O$ will therefore measure the time $d\hat{t}_O$ corresponding to the coordinate time interval (inverse frequency) dt as

$$d\hat{t}_O = -u_O^0 \, g_{00}(r_O) \, dt = d\tau$$
$$= \sqrt{-g_{00}} \, dt \simeq \left(1 - \frac{GM}{r_O}\right) dt, \tag{2.42}$$

In the last equality we have quoted the result only to first order in the local gravitational potential $U_O = -GM/r_O$, as we do from now on. So as noted above, the observer fixed with respect to the frequency source measures a time interval $d\hat{t}_O = d\tau$, the proper time.

This fixed observer is not, however, freely-falling and we may also want to consider measurements made in the LIFs corresponding to such observers, whose worldlines satisfy the geodesic equation. Consider therefore an observer (A) freely-falling radially along a trajectory with boundary condition $dr/dt = 0$ at $r = \infty$. Solving the geodesic equation, we find this observer has normalised 4-velocity $u_A^\mu = \left((1 - 2GM/r_O)^{-1}, -(2GM/r_O)^{1/2}, 0, 0\right)$ at $r = r_O$. So observer A measures the time interval considered above as

$$d\hat{t}_A = -u_A^0 \, g_{00}(r_O) \, dt = dt \ . \tag{2.43}$$

Alternatively, consider the freely-falling observer (B) in a circular orbit with constant radius r_O. This observer has 4-velocity $u_B^\mu = (dt/d\sigma) \, (1, 0, 0, \omega)$, where the angular frequency $\omega = d\phi/dt$ and the orbital velocity is $v = \omega r = (GM/r)^{1/2}$. It follows from (2.40) that $dt/d\sigma = (1 - 3GM/r_O)^{-1/2}$. So observer B measures the time interval as

$$d\hat{t}_B = -u_B^0 \, g_{00}(r_O) \, dt \simeq \left(1 - \frac{1}{2}\frac{GM}{r_O}\right) dt \ . \tag{2.44}$$

Now, since the coordinate time dt is not a measured quantity, what is important here is only the *ratios* of measurements amongst the different observers. Every time measurement is only a ratio with respect to another clock, which is also affected by the spacetime curvature. So here, we find to leading order in GM/r_O,

$$d\hat{t}_A = \left(1 + \frac{GM}{r_O}\right) d\hat{t}_O, \qquad d\hat{t}_B = \left(1 + \frac{1}{2}\frac{GM}{r_O}\right) d\hat{t}_O \ . \tag{2.45}$$

In both cases, the freely-falling observers, who are moving relative to the fixed location of the frequency source, measure a *greater* time interval than the fixed observer. Comparing with (2.41), we can check directly that these results may

---

So, as required by the relativity principle, we indeed recover the same time dilation factor as above.

be interpreted as the local special relativistic time dilation factor $\gamma(v^2)$, where $v^2 = g_{ij} \frac{dx^i}{d\tau} \frac{dx^j}{d\tau}$ is the appropriate squared physical velocity measured with the fixed observer's time. It is perhaps worth emphasising that the different freely-falling observers do measure different times.

Note also that since the ratio of these time measurements $d\hat{t}_O$, $d\hat{t}_A$ and $d\hat{t}_B$ is simply determined by the ratio of the relevant 4-velocity components u_O^0, u_A^0 and u_B^0, with the metric factor $g_{00}(r_O)$ cancelling, these results illustrate an important feature of GR, viz. that purely *local* measurements are not dependent on the absolute value of the gravitational potential.[11]

These examples involve observers at the same point in the gravitational potential. Considering observers at different heights gives rise to the *gravitational redshift* effect. To derive this, consider a light wave emitted from a source (E) at $r = r_E$ radially upwards to the observer (O) at $r_O = r_E + h$. It is straightforward to see from the null geodesic equation that the coordinate time dt between successive wave maxima is the same at the receiver (O) as at the emitter (E).[12] Using the results above, we see that the period of the wave measured by an observer E fixed at the location of the emitter is therefore $d\hat{t}_E = (1 - GM/r_E)\, dt$, while measured at the receiver O it is $d\hat{t}_O = (1 - GM/r_O)\, dt$. The ratio of observed frequencies is therefore

$$\frac{\nu_O}{\nu_E} = \frac{d\hat{t}_E}{d\hat{t}_O} = \left(1 - \frac{GM}{r_E}\right)\left(1 - \frac{GM}{r_O}\right)^{-1}$$
$$\simeq 1 - \frac{GMh}{r_E^2}, \tag{2.46}$$

for $h \ll r_E$. The observer at height h above the emitter therefore measures the light *redshifted* relative to the measurement at the emitter, by a factor $\Delta\nu/\nu \simeq GMh/r_E^2$. Notice that for these fixed observers at different spacetime points, the measured effect

---

[11]The local flatness of Riemannian spacetime means that at each point we can construct Riemann normal coordinates in which the metric has the form

$$g_{\mu\nu}(x) = \eta_{\mu\nu} + \frac{1}{3} R_{\mu\rho\nu\sigma} x^\rho x^\sigma + O(x^3),$$

since $\Gamma_{\mu\nu}^\lambda \sim 0$ in these coordinates. The curvature tensor involves the second derivatives of the metric (and hence the potential $U(r) = -GM/r$). It follows that physical measurements in a local laboratory of size ℓ only depend on the curvature at order $O(|U|\ell^2/L^2)$ where L is the curvature length scale, which is hugely suppressed relative to the potential itself (see e.g. [62] for a recent comment).

[12]For a photon following a null geodesic, its trajectory is characterised by $g_{00}dt^2 + g_{rr}dr^2 = 0$, so the coordinate time to travel from E to O is simply

$$t_O - t_E = \int_{r_E}^{r_O} dr \left(1 - \frac{2GM}{r}\right)^{-1},$$

where, since the metric is static, the integral on the r.h.s. is a function of the space coordinates only. The time of flight for successive maxima is therefore the same, so the coordinate wave period dt is the same at the receiver and emitter.

depends only on the *difference* of the gravitational potentials, not their absolute values, and is therefore suppressed by an additional factor h/r_E. This was first observed experimentally in the famous Pound–Rebka experiment [63] using gamma rays and exploiting the Mössbauer effect.

Another interesting gravitational redshift test may be made using atom matter-wave interferometry [64, 65]. To see the principle involved, consider two identical frequency sources (ideal clocks) P and Q at positions r_P and r_Q with velocities v_P and v_Q respectively. A simple calculation combining (2.46) with (2.39) and (2.41) shows that the difference in frequencies associated with P and Q as measured in the laboratory frame is given by

$$\frac{\Delta \nu_{P-Q}}{\nu} \equiv \frac{\nu_P - \nu_Q}{\nu} = U(r_P) - U(r_Q) - \frac{1}{2} \left(v_P^2 - v_Q^2 \right), \qquad (2.47)$$

to leading order in the weak field, low-velocity approximation. At this order we can simply take the denominator factor ν to be the common flat-spacetime limit of the source frequencies.

Now, if P and Q start from a common position at $t = 0$ and follow different trajectories before being brought back to a common point at time T, they will acquire a phase difference given by

$$\Delta \phi = \int dt \, \Delta \omega_{P-Q}$$
$$= \omega \int_0^T dt \left[g \left(r_P - r_Q \right) - \frac{1}{2} \left(v_P^2 - v_Q^2 \right) \right], \qquad (2.48)$$

where ω is the angular frequency and $g = GM/R^2$ is the Earth's gravitational acceleration.

Notice that this result can be derived[13] very elegantly as

---

[13]Explicitly,

$$\int d\sigma_P = \int dt \left[-\left(g_{00}(P) + g_{ij}(P) v_P^i v_P^j \right) \right]^{1/2}$$
$$= \int dt \left(-g_{00}(P) \right)^{1/2} \left(1 + g_{ij}(P) v_P^i v_P^j / g_{00}(P) \right)^{1/2}$$
$$\simeq \int dt \left(1 - \frac{GM}{r_P} - \frac{1}{2} v_P^2 \right),$$

to first order in the small quantities GM/r_P and v_P^2. The result (2.48) follows immediately as

$$\Delta \phi = \omega \int \left(d\sigma_P - d\sigma_Q \right)$$
$$= \omega \int dt \left[U(r_P) - U(r_Q) - \frac{1}{2} \left(v_P^2 - v_Q^2 \right) \right].$$

$$\Delta\phi = \omega \int \left(d\sigma_P - d\sigma_Q\right) . \tag{2.49}$$

That is, the phase difference measures the difference in the proper times along the distinct trajectories of P and Q.

In the atom interferometry experiment [64, 65], a laser cooled atom projected vertically is subjected to three pulses from a pair of crossed laser beams. The first puts the atom into a superposition of quantum states with different momenta, so they follow different trajectories. The second pulse reverses their spatial separation and brings them back to a common point at which the third laser pulse records the phase difference of the superposed states. The matter waves oscillate with the Compton frequency $\omega_C = mc^2/\hbar$.

The experiment can be arranged so that the gravitational redshift contribution to the phase difference (2.48) can be isolated, leaving

$$\Delta\phi_{redshift} = \omega_C \int_0^T dt \, g \, \Delta r(t), \tag{2.50}$$

where $\Delta r(t)$ is the vertical separation of the trajectories and we continue to set $c = 1$.

This experiment allows a very high precision test of WEPc in the laboratory. Further details and prospects for repeating it with antihydrogen in order to test WEPc directly in a neutral, pure antimatter system will be discussed in Sect. 3.3.

This concludes our brief summary of the fundamental principles of GR, together with some of the basic theory of time meaurements which we will need later in describing the effect of curved spacetime on atomic spectroscopy measurements in the Earth's gravitational field and in orbits around the Earth or Sun. In the next section, we discuss some implications of any possible violation of the predictions of GR, in particular any anomalous gravitational measurements which would distinguish between matter and antimatter.

2.5 Breaking General Relativity and the Equivalence Principles

As we have seen, GR provides an elegant and compelling view of gravity whose predictions, however radical and counter-intuitive, have passed all experimental tests for more than a century. Nevertheless, as a purely classical theory, we know that eventually GR must be completed into a full quantum theory of gravity, and we may expect that even at the level of a low-energy effective Lagrangian small deviations from the simplest formulation of GR could arise. While such modifications of the dynamics of GR are relatively easy to implement, as we shall see what is not so straightforward is to envisage changes to the theory that distinguish matter and antimatter in a way that could be observable in the current generation of antimatter experiments. An interesting discussion of the fundamental issues involved is given in [66].

A conservative approach is to start by retaining the core principles of GR—the description of gravity as a Riemannian manifold and local Lorentz invariance—but modifying the dynamics described by the action (2.29).

The action (2.29) was determined using the criterion that the matter couplings to gravity are through the connection only, not the curvature. This means that in a LIF, the equations of motion take their special relativistic form, thereby implementing the Strong Equivalence Principle. The simplest extension of GR is therefore to violate the SEP by including explicit curvature terms in the action. These transform covariantly so we maintain local Lorentz invariance, but the *dynamics* is now modified to be sensitive to the curvature at each spacetime point.

For example, for the free Dirac theory we can generalise the action to:

$$S = \int d^4 x \sqrt{-g} \left(\frac{R}{16\pi G} + \bar{\psi} \left(i\gamma^\mu D_\mu - m \right) \psi \right.$$
$$+ e R \bar{\psi}\psi + f \bar{\psi} D^2 \psi + a \, \partial_\mu R \, \bar{\psi}\gamma^\mu\psi + b R \, \bar{\psi} i\gamma^\mu \overleftrightarrow{D}_\mu \psi$$
$$\left. + c R_{\mu\nu} \bar{\psi} i\gamma^\mu \overleftrightarrow{D}^\nu \psi + d D_\nu \bar{\psi} i\gamma^\mu \overleftrightarrow{D}_\mu D^\nu \psi + \cdots \right), \tag{2.51}$$

with the obvious notation $\gamma^\mu = \gamma^a e_a{}^\mu$. Here, in the spirit of regarding (2.51) as a low-energy effective Lagrangian for some more fundamental UV complete theory, we have included all the operators of dimension 4, 5 and 6 which are quadratic in the spinor field and, for ease of illustration only, conserve parity. The coefficients $a, \ldots, f$ carry the appropriate inverse dimensions of mass, potentially set by the scale of the UV completion.

A notable feature of this SEP-violating effective Lagrangian for fermions is that it does not involve the Riemann curvature tensor $R_{\mu\rho\nu\sigma}$ itself, only the contracted forms—the Ricci tensor $R_{\mu\nu}$ and scalar R. A general discussion of such SEP-violating actions is given in [59, 67].

This is not true of the extension to QED, since there are explicit SEP-violating couplings of the photon field strength to the Riemann curvature:

$$S = \int d^4 x \sqrt{-g} \left(-\tfrac{1}{4} F_{\mu\nu} F^{\mu\nu} + \tilde{a} R F_{\mu\nu} F^{\mu\nu} + \tilde{b} R_{\mu\nu} F^{\mu\rho} F^\nu{}_\rho \right.$$
$$\left. + \tilde{c} R_{\mu\rho\nu\sigma} F^{\mu\nu} F^{\rho\sigma} + \tilde{d} D_\mu F^{\mu\rho} D_\nu F^\nu{}_\rho + \cdots \right). \tag{2.52}$$

In the same spirit, we can also modify the purely gravitational dynamics by adding terms of quadratic or higher order in the curvature, e.g.

$$S = \int d^4 x \sqrt{-g} \left(\hat{a} R^2 + \hat{b} R_{\mu\nu} R^{\mu\nu} + \hat{c} R_{\mu\rho\nu\sigma} R^{\mu\rho\nu\sigma} + \cdots \right). \tag{2.53}$$

There is a large literature on modified (higher-derivative) gravity theories of this type, motivated in part by superstring theory.

The extra terms in (2.51) and (2.52) modify the energy-momentum tensor and the equations of motion for the fields, which no longer necessarily follow geodesics. These couplings may be different for different particles, and may in principle even distinguish matter and antimatter. So in general, the universality of free-fall (WEPff) is lost in this GR extension.

On the other hand, it is not so clear how this modified dynamics would affect the gravitational time dilation results described in the last section. These depend purely on the nature of the metric and spacetime, and since all matter is still coupled to the same metric the universality of clocks (WEPc) seems at first sight to be maintained. However, we now have to consider the possibility of these new gravitational interactions dynamically modifying the spectrum of anti-atoms, for example, inducing physical differences between time/frequency measurements with different atomic/anti-atomic clocks.

The SEP-violating QED action is in some respects analogous to the Lorentz-violating action considered in Sect. 2.3 and the phenomenological consequences may in some cases be similar. There are, however, many important differences, notably that the rôle of the Lorentz-violating couplings is played here by the covariant curvature tensors. This particularly affects the C, P and T character of the operators. These can be read off from Table 2.1, remembering that the spinor quantities are all referred to the local Minkowski frame using the implicit vierbein. A careful analysis is given in [59]. Importantly, all these operators are CPT invariant.

A particularly interesting example is the term

$$\mathcal{L}_a = a\, \partial_\mu R\, \bar{\psi}\gamma^\mu\psi \,. \tag{2.54}$$

This is C and T odd, P even, and therefore CP odd, CPT even. The fact that, uniquely amongst those in (2.51), this operator is CP odd, allows it to modify the propagation (dispersion relations) for fermions and antifermions *differently*.[14] We can therefore conclude that in this SEP-violating GR extension, there is an induced matter-antimatter asymmetry in a background gravitational field with a time-varying Ricci scalar [43, 59].

This is a radical consequence of what is a natural and relatively mild modification of GR, in which the core geometrical structure is maintained while the dynamics is extended to include direct curvature couplings. That said, the theoretical consistency of the theory based on (2.51), (2.52) needs to be analysed critically. Just as in the Lorentz-violating QED extension, if the action is viewed as a fundamental theory in itself it is at risk of violating causality—indeed (2.52) implies birefringent superluminal propagation of photons. Once more, we are therefore led to the point of

---

[14]The divergence of the Ricci scalar $\partial_\mu R$ acts here as a background vector field coupling to the current $J^\mu = \bar{\psi}\gamma^\mu\psi$, so the coupling depends on the corresponding charge and is equal and opposite for particles and antiparticles. We return to this idea in Sect. 2.6 when we consider extra background fields, such as a "gravivector", which could have gravitational strength interactions distinguishing matter and antimatter.

view of regarding (2.51), (2.52) as a low-energy effective Lagrangian whose intrinsic inconsistencies may be modified by a suitable UV completion [53, 54].

Now, it is a remarkable fact that even if we start from the usual QED action (2.29) in curved spacetime, loop corrections to the electron and photon propagators generate precisely the operators in the effective action (2.51), (2.52) and (2.53), with the exception of the CP-violating operator.[15] This means that conventional QFT in curved spacetime *automatically* violates the SEP at low energies due to radiative quantum corrections, while maintaining causality and unitarity [55–58]. Furthermore, in a BSM theory with CP-violating couplings, even the operator (2.54) is generated. This provides the theoretical basis for a recently proposed mechanism— radiatively-induced gravitational leptogenesis—for generating the observed matter-antimatter asymmetry of the universe [43, 68, 69].

We therefore see that in principle it is perfectly possible for nature to be described by a SEP-violating low-energy effective Lagrangian, exhibiting WEPff violation and perhaps also independently WEPc violation, while maintaining the fundamental principles of local QFT and GR.

Unfortunately, there is an obvious problem as far as terrestrial antimatter experiments are concerned, viz. that the Schwarzschild spacetime describing the gravitational field on the Earth's surface is Ricci flat ($R_{\mu\nu} = 0$, $R = 0$), although the Riemann tensor itself is non-vanishing. The modified Dirac action (2.51) is therefore insensitive to the Earth's gravitational field, although the photon action (2.52) and light propagation is affected. The only way out would be for an extended pure gravity action to generate a modified solution with non-vanishing Ricci tensor which could couple to fermions.

Nevertheless, this is only one approach to extending GR and violating the equivalence principles. More radically, we could consider adding new fields and interactions, discussed briefly in the following section, or even modify the metric structure of spacetime and consider alternatives to the simple Riemannian picture.

In the spirit of the latter approach, we could consider bimetric, or multi-metric, theories where different fields couple to different metrics imposed on the background Riemannian manifold. An interesting gedanken experiment is then to reconsider the gravitational redshift scenario of the previous section, but where the emitter and receiver couple to different metrics. A simple case to illustrate the idea would be if the receiver, but not the emitter, was described by an effective Lagrangian of the form (2.51) with the coefficient $c \neq 0$. This would correspond to an *effective metric*:

$$g_{\mu\nu}^{\text{eff}} = g_{\mu\nu} + 2c R_{\mu\nu} . \qquad (2.55)$$

---

[15]The scale of these quantum effects is $O(\alpha\lambda_c^2/L^2)$, where α is the fine structure constant, $\lambda_c = 1/m$ is the Compton wavelength of the virtual particles arising in the self-energy or vacuum polarisation Feynman diagrams, and L is the curvature scale. Of course, these quantum effects only become significant in an epoch of extremely high curvatures when the quantum scale becomes comparable with the curvature scale. In the BSM model of [43, 68, 69], this is set by the heavy neutrino mass. Remarkably, in the high curvature and temperature regime of the early universe, this mechanism may generate a sufficient difference in the dispersion relations of the light neutrinos and antineutrinos to give rise to the current observed baryon-to-photon ratio of $O(10^{-10})$.

This would give an additional curvature-dependent contribution to the redshift formula (2.46), violating WEPc. Unfortunately, this particular operator is CP conserving and we do not have a model which would differentiate matter from antimatter in this simple way.

While such curvature-dependent modifications to an effective metric are relatively easy to contemplate within the worldview of GR, this is not true of the totally phenomenological approach used especially in some of the early literature on equivalence principle violation in antimatter systems (see e.g. [70]). These imagine an effective metric with g_{00} component of the form

$$(g_{00})_{\text{eff}} = 1 - \alpha_g \frac{2GM}{r}, \tag{2.56}$$

where α_g is a parameter to be constrained by experiment, independently for matter and antimatter. Clearly this directly enters the redshift formula. It is not clear, however, that this is a useful parametrisation. In the first place, it is very hard to imagine a causal extension of GR in which $\alpha_g - 1$ is non-zero and different for particles and antiparticles. Moreover, this would imply that local experiments would be sensitive not just to the local curvature but to the absolute value of the gravitational potential $U(r) = -GM/r$ itself. Yet this becomes greater for ever more distant astronomical structures, ranging from the Sun to the Galaxy or even the Virgo Cluster. Numerical values are discussed later in Sect. 3.3.2, but for now we simply note that the loss of universality implied by different non-vanishing α_g parameters for different fields would therefore not only violate the equivalence principle but would involve a rather bizarre non-locality of matter-gravity interactions observed on Earth. This parametrisation is, however, still quite widely used to quantify measurements in experiments testing the equivalence principle. We return to this issue in Sect. 3.3.

2.6 'Fifth' Forces, $B - L$ and Supergravity

This brings us to the possibility that any anomalous results that may be seen in future antimatter experiments are due not to a modification of the fundamental principles of our existing theories but due to new interactions, so-called 'fifth forces', beyond the standard model (BSM).

Since our focus is on phenomena which would distinguish matter and antimatter, we are led to consider new interactions that are mediated by the exchange of a spin 1, or vector, boson. Whereas spin 2 bosons (such as the graviton, corresponding to a tensor field) and spin 0 bosons (scalars, such as the Higgs) couple equally to particles and antiparticles, spin 1 bosons couple to a vector current with a strength proportional to the corresponding charge. Just as in QED with photon exchange, this charge is opposite for particles and their antiparticles.

It is then immediately apparent that while they would distinguish matter and antimatter, such new vector interactions would necessarily affect the dynamics of purely matter systems. They are therefore already highly constrained by comparisons of ordinary matter interactions with conventional theories, irrespective of future tests with antimatter.

Here, we consider especially two theoretically well-motivated classes of theory involving new vector interactions. The first is a BSM theory with gauged $B - L$ symmetry, which contains an additional Z' boson compared to the standard model, as well as right-handed neutrinos. The second is $\mathcal{N} \geq 2$ supergravity, in which the graviton lies in a supersymmetry multiplet with a spin 1 '*gravivector*' boson.

2.6.1 Gauged $B - L$ Theories

A particularly compelling class of BSM theories which could produce long-range forces distinguishing matter and antimatter are those where the $B - L$ symmetry of the standard model is gauged. The full gauge group is extended to $SU(3)_C \times SU(2)_L \times U(1)_Y \times U(1)_{B-L}$, with the introduction of a new abelian gauge field and corresponding spin 1 boson Z'. We denote the new gauge coupling by g' and the corresponding 'fine structure constant' $\alpha' = g'^2/4\pi$.

To understand the theoretical motivation, we need to briefly review some key ideas involving anomalies in QFT. An anomaly arises when a symmetry which is exact in the classical theory is broken in the full quantum theory. There are many ways to understand the origin of anomalies, which fundamentally involve the deep mathematical structure of gauge theories, but in the simplest perturbative picture they arise from the behaviour of 3-point Feynman diagrams where the vertices are currents $J_\mu^a = \bar{\psi}\gamma_\mu T^a \psi$, for some symmetry generators T^a, and left or right-handed fermions $\psi_{L,R}$ go round the triangle; see Fig. 2.3. An otherwise conserved current J_μ^a will be anomalous if the coefficient $\mathcal{A}$ in the triangle diagram is non-zero, where

$$\mathcal{A} = \sum_{L \text{ reps}} T^a \{T^b, T^c\} - \sum_{R \text{ reps}} T^a \{T^b, T^c\}, \tag{2.57}$$

and the sums are over the representations of the L and R-handed fermions in the corresponding symmetry groups.[16]

If these anomalies affect only *global* currents, such as the axial current in QED where the anomaly determines the physical decay $\pi^0 \to \gamma\gamma$, the quantum theory remains consistent. However, if the anomalies affect *local* currents, i.e. if a gauge field is coupled to an anomalous current, then *unitarity* will be broken in the full QFT. Any realistic particle physics theory must therefore only involve gauge fields which couple to anomaly-free conserved currents. Clearly, this places important constraints on the fermion content of the theory.

---

[16]This notation is over-simplified for clarity—the generators T^a, T^b, T^c in (2.57) may refer to different symmetry groups.

Fig. 2.3 Feynman diagram
for a 3-current Green
function $\langle 0| J_\mu^a \; J_\nu^b \; J_\lambda^c |0 \rangle$ with
L and R-handed fermions in
the loop. Gauge fields
A_μ^a, A_ν^b and A_λ^c couple to the
currents if the corresponding
symmetries are gauged. This
diagram may contribute an
anomaly which breaks
conservation of the currents

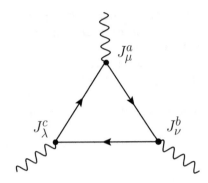

In the standard model, both baryon number B and lepton number L are conserved
at the classical level but are anomalous in the full quantum theory. However, the
anomaly $\mathcal{A}$ cancels when one vertex is the combined $B - L$ current while the others
refer to the gauge symmetries of the standard model. In fact, $B - L$ is the only
anomaly-free conserved global symmetry in the standard model.

This provides the motivation to extend the standard model gauge group to include
$U(1)_{B-L}$. However, unitarity now demands that all triangle diagrams with one or
more $U(1)_{B-L}$ vertices must have a vanishing anomaly coefficient. In particular, for
the $U(1)_{B-L}^3$ triangle, the anomaly is

$$
\begin{aligned}
\mathcal{A} &= \sum_{L \text{ reps}} Q_{B-L}^3 - \sum_{R \text{ reps}} Q_{B-L}^3 \\
&= \left(3 . 2 \left(\frac{1}{3} \right) + 2 \, (-1) \right) - \left(3 \left(\frac{1}{3} \right) + 3 \left(\frac{1}{3} \right) + (-1) \right) \\
&= -1,
\end{aligned}
\tag{2.58}
$$

where we have shown the $B - L$ charges and $SU(3)_C$, $SU(2)_L$ multiplets for one
generation of quarks and leptons. To obtain an anomaly-free theory with $\mathcal{A} = 0$ we
must therefore add a new R-handed neutrino for each flavour generation, so the quark
and lepton representations are precisely balanced.

To summarise, the fundamental principle of *unitarity* requires that the standard
model extension with $U(1)_{B-L}$ gauged requires three right-handed neutrinos as well
as the new gauge boson Z'. Of course, this is a very welcome bonus as it permits
neutrinos to acquire non-vanishing masses as required by experiment (in contrast to
the standard model itself where only L-handed neutrinos and R-handed antineutrinos
exist and are therefore required to be massless).

We now focus on the gauge sector and the new Z' boson. There are two distinct
realisations of the theory, one with unbroken $U(1)_{B-L}$ and the other where $U(1)_{B-L}$
is spontaneously broken through the interaction with a new Higgs field.

(i) *Unbroken $U(1)_{B-L}$*

In this model [71], the L and R-handed neutrinos combine into three flavours of massive Dirac neutrinos with masses determined, as for the other leptons, by arbitrary Yukawa couplings to the $SU(2)_L$ doublet Higgs field ϕ, that is

$$\mathcal{L}_D = y_D^{ij} \left(\overline{\ell_L^i} \, \tilde{\phi} \, \nu_R^j + \text{h.c.} \right), \tag{2.59}$$

in standard notation, for generations $i = 1, 2, 3$. Since the Higgs has zero $B - L$ charge, its VEV does not break the $U(1)_{B-L}$ gauge symmetry and at first sight it appears the Z' must remain massless.

However, for abelian gauge theories (only), there is an alternative to the Higgs mechanism which can give a mass to the Z'. This is the Stueckelberg mechanism. It involves the introduction of a scalar field analogous to a conventional Higgs field but where its modulus (corresponding to the physical Higgs boson) is non-dynamical while its phase (corresponding to the Goldstone boson) provides the extra degree of freedom required to give a massive spin 1 boson. Essentially, it is a gauged $U(1)$ non-linear sigma model, in contrast to the Higgs theory which is a gauged linear sigma model.

The details, and the proof of renormalisability and unitarity—which work only for a $U(1)$ gauge theory—need not concern us here (see Ref. [72]). The essential point is simply that there exists a theoretically consistent model in which the $U(1)_{B-L}$ gauge symmetry remains unbroken while the Z' boson acquires a mass, which is unconstrained by theory. In this model, the light neutrinos are of Dirac type.

(ii) *Spontaneously broken $U(1)_{B-L}$*

In this case (see for example [73]), we introduce a new complex $SU(2)_L$ singlet Higgs field Φ with $B - L$ charge $Q_{B-L} = -2$. If this acquires a VEV, the Z' boson acquires a mass in the usual way and there remain two physical Higgs bosons in the spectrum. The Φ field can couple to the R-handed neutrino fields, giving them a Majorana mass proportional to its VEV in addition to the Dirac mass (2.59). The interaction term in the Lagrangian is

$$\mathcal{L}_M = y_M^{ij} \left(\overline{\left(\nu_R^i \right)^c} \, \nu_R^j \, \Phi + \text{h.c.} \right). \tag{2.60}$$

This interaction allows explicitly $B - L$ violating processes involving the R-handed neutrinos, such as neutrinoless double beta decay. The spectrum comprises two massive Majorana neutrinos for each fermion generation.

What this discussion illustrates is that it is highly non-trivial to add a new 'fifth force' interaction to the standard model, even one as seemingly innocuous as a coupling to baryon or lepton number. There are fundamental principles, in this case unitarity (via anomaly freedom), which severely constrain the particle content and interactions on purely theoretical grounds.

The phenomenology of these two variants of gauged $U(1)_{B-L}$ theory therefore differs in key aspects but both have been extensively studied, especially in the parameter range of most interest to particle physicists exploring neutrino phenomenology and the implications of a potential new Z' boson in the mass range of interest at the LHC [73].

Here, we are more interested in two mass windows for the Z' which could impact on low-energy antimatter experiments. The first is $m_{Z'} \sim 1$ keV, which implies a short-range interaction with range $\lambda = 1/m_{Z'} \sim 10^{-10}$ m, sufficient to influence the internal structure of atomic systems.

The second is $m_{Z'} \lesssim 10^{-14}$ eV, corresponding to a new long-range interaction with range $\lambda \gtrsim 10^7$ m. This (including the massless limit $m_{Z'} = 0$) would imply a macroscopic interaction between the Earth, which has a huge $B - L$ charge, and a single atom or molecule with non-vanishing $B - L$ at the surface. This force is in principle measurable in free-fall experiments of the type which may be realised for suitable antimatter systems at the CERN AD.

Existing constraints on the strength of a new $U(1)_{B-L}$ interaction over the entire range of Z' mass from zero to GeV/TeV values are summarised in [71]. For $m'_Z \sim 1$ keV, limits arise from low-energy electron neutrino scattering, particularly elastic scattering of solar neutrinos, which (for the unbroken theory) gives $\alpha' \lesssim 10^{-13}$, and from stellar astrophysics, which gives a more stringent bound $\alpha' \lesssim 10^{-31}$ from an analysis of energy-loss mechanisms involving the Z' and new neutrinos. See Fig. 3 of [71] for a clear compilation of the experimental bounds.

For a very light Z', existing equivalence principle experiments on ordinary matter, especially highly precise torsion balance experiments, constrain the $U(1)_{B-L}$ gauge coupling to the range $\alpha' \lesssim 10^{-49}$. The extremely small value arises because of the large $B - L$ charge of the Earth, equal to the number of neutrons, which is of order $Q_{B-L}^{\text{Earth}} \sim 10^{51}$. Such long-range forces are generally parameterised in the form of a modified gravitational potential derived from single-boson Born exchange,

$$
\begin{aligned}
V(r) &= -\frac{G_\infty m_1 m_2}{r} + \frac{\alpha' Q_{B-L}^1 Q_{B-L}^2}{r} e^{-m_{Z'} r} \\
&= -\frac{G_\infty m_1 m_2}{r} \left(1 - \tilde{\alpha}\, e^{-r/\lambda} \right),
\end{aligned}
\tag{2.61}
$$

where $\tilde{\alpha} = \alpha' Q_{B-L}^1 Q_{B-L}^2 / G_\infty m_1 m_2$. Note that we have introduced the 'fundamental' Newton constant G_∞ here, since an infinite range fifth force would change the effective constant G_N measured in experiment. The same ambiguity would affect the definition of the Planck mass from $G = 1/M_{pl}^2$. Also note crucially that the Z' exchange force is *attractive* when the $B - L$ charges of the interacting bodies are *opposite*, and repulsive when they are the same, in exact analogy with the electric force between charged particles.

2.6.2 $\mathcal{N} \geq 2$ supergravity

Another class of theories which naturally involve extra vector boson interactions are supergravities with $\mathcal{N} \geq 2$. Supersymmetry is an extension of the Poincaré symmetry of flat spacetime to include new generators $\mathcal{Q}_\alpha^a$, where α is a spinor index and $a = 1, \ldots \mathcal{N}$ counts the number of supersymmetries, which satisfy *anticommutation* relations. This extended spacetime symmetry is known as a graded Lie algebra. If supersymmetry is promoted to be a *local*, rather than global, symmetry, the resulting quantum field theory is supergravity.

The impact of supersymmetry on the spectrum of a QFT is to relate fields, or particles, with spins differing by $1/2$. So the simplest $\mathcal{N} = 1$ supergravity has one spin 2 graviton and one spin $3/2$ Majorana gravitino. These supermultiplets become progressively longer as the number of supersymmetries is increased. $\mathcal{N} = 2$ supergravity has one graviton, two gravitinos, and one spin 1 *gravivector* (all massless, and balancing $2 + 2 = 4$ bosonic and $2 \times 2 = 4$ fermionic degrees of freedom). Ultimately we arrive at $\mathcal{N} = 8$ supergravity, with a spectrum comprising 1 spin 2, 8 spin $3/2$, 28 spin 1, 56 spin 1/2 and 70 spin 0 massles particles (128 bosonic and 128 fermionic degrees of freedom). Theories with $\mathcal{N} > 8$ necessarily involve fields with spins greater than 2 and are not evidently unitary.

For our purposes, we are interested in the potential physical effects of a gravivector on low-energy antimatter experiments. The simplest theory to consider is $\mathcal{N} = 2$ supergravity in which (setting aside the gravitinos for the moment) the gravitational force due to the exchange of the massless graviton is complemented by a new 'fifth force' mediated by gravivector exchange. Matter fields describing the standard model fermions must be introduced into the theory separately as $\mathcal{N} = 2$ supermultiplets with mass m, comprising two Majorana fermions χ^i and two complex scalar fields.

The construction of supergravity Lagrangians is highly technical and the details need not concern us here. The key point is that the supersymmetry algebra relates the matter coupling of the gravivector to that of the graviton.[17] Letting $\kappa^2 = 4\pi G$, the coupling of the gravivector A_μ to the massive Dirac fermion $\psi = \frac{1}{\sqrt{2}} \left(\chi^1 + i\chi^2 \right)$ in the matter supermultiplet is of the familiar form,

$$\mathcal{L}_{\text{int}} = i g' \bar{\psi} \gamma^\mu \psi \, A_\mu, \tag{2.62}$$

with $g' = \kappa m$, that is $\alpha' = m^2 / M_{pl}^2$.

From this point, the analysis mirrors that presented above for the $U(1)_{B-L}$ gauge theory, with the gravivector boson playing the rôle of the Z'. The gravivector is massless in the theory with unbroken supersymmetry, but may be given a mass

---

[17]Briefly, the global $\mathcal{N} = 2$ supersymmetry algebra acting on the massive matter supermultiplets contains a central charge, which is matched to a gauge transformation on the gravivector field induced by the local $\mathcal{N} = 2$ supersymmetry transformation on the graviton supermultiplet. This identification of the gravivector as the gauge boson for the $\mathcal{N} = 2$ central charge fixes its coupling to the matter supermultiplet. The full Lagrangian for $\mathcal{N} = 2$ supergravity coupled to a massive matter supermultiplet was first derived explicitly in [74, 75].

m_V (along with the gravitinos) by breaking supersymmetry through a super-Higgs mechanism. By evaluating the 1-graviton and 1-gravivector Born exchange diagrams, we deduce the following effective potential between matter/antimatter fermions,

$$V(r) = -\frac{Gm_1m_2}{r} + \frac{g_1g_2}{4\pi r} e^{-m_Vr}, \qquad (2.63)$$

with $g = \pm\kappa m$ for a fermion (antifermion) respectively.

Rewriting (2.63) and substituting for the couplings g_1, g_2, we find the remarkable result [75–77],

$$V(r) = -\frac{Gm_1m_2}{r} \left(1 \mp e^{-m_Vr}\right), \qquad (2.64)$$

where the $-$ sign gives the potential between two fermions or two antifermions, with the $+$ sign for a fermion-antifermion interaction. With a massless gravivector, there is therefore an *exact cancellation* between the usual gravitational force and the new gravivector-induced interaction between pairs of elementary fermions or pairs of antifermions. In this sense, the gravivector force is 'antigravity'. Notice, however, that this is exactly the opposite of the popular use of 'antigravity', which speculates (without foundation) about a repulsive gravitational interaction between matter and antimatter. For a fermion-antifermion pair, the gravivector force is attractive and doubles the strength of the usual gravitational attraction.

A similar effect occurs in higher supergravities. For example, in $\mathcal{N} = 8$ super-gravity there are contributions from the graviton, gravivector and spin 0 graviscalar fields. The spin 2 and spin 0 exchange diagrams are always attractive, while the spin 1 exchange may be attractive or repulsive. For unbroken $\mathcal{N} = 8$ supergravity, the coupling is $g = 2\kappa m$ and the contributions again cancel exactly for a fermion-fermion or antifermion-antifermion interaction [77].

For long-range interactions, the gravivector is therefore constrained by the same experimental tests of the weak equivalence principle (WEPff) on ordinary matter that constrain the Z' interaction in $U(1)_{B-L}$ gauge theory. However, in supergravity, as we have seen, the coupling $\alpha' = m^2/M_{pl}^2$ is *fixed*, so the only free parameter is the gravivector mass m_V.[18]

An additional complication in applying the potential (2.64) to matter comprised of nucleons is that the fundamental interaction is between the gravivector and the elementary quarks in the bound-state nucleon, whereas the graviton couples to the full energy-momentum tensor including the gluons (the gravivector-gluon coupling vanishes). Since the quark masses constitute approximately 1 percent of the nucleon mass, there is scope for the gravivector and graviton interactions with nucleons to differ by a factor of $O(10^{-2})$. A comprehensive account of the experimental limits on gravivector (and graviscalar) interactions in $\mathcal{N} = 2$ and $\mathcal{N} = 8$ supergravity from

---

[18]Using a light quark mass, the gravivector coupling is of order $\alpha' = m_q^2/M_{pl}^2 \sim 10^{-42}$. Comparing this with the above limit $\alpha' \lesssim 10^{-49}$ on the coupling of a light Z' boson with range greater than the Earth's radius, shows immediately that in this theory a massless or light gravivector of this range is already ruled out by existing equivalence principle tests on matter.

equivalence principle tests on matter, including these considerations, is given in [78]. This concludes that the range of the gravivector interaction is bounded by $\lambda \lesssim 1$ m, with the gravivector mass $m_V \gtrsim 10^{-6}$ eV.

The conclusion is that while the strength of the gravivector interaction is fixed by supersymmetry to be of the same order as the gravitational force, its range is already experimentally constrained *in these theories* to be too short to have any appreciable effect on single-atom free-fall experiments dependent on the mass of the Earth, such as the WEPff tests proposed in antimatter experiments.

2.6.3 S, V, T Background Fields

Although these supergravities are not themselves realistic BSM theories encompassing the standard model, they do suggest we investigate the experimental consequences of potential gravivector and graviscalar interactions in more generality. In this section, we therefore adopt a purely phenomenological approach and consider the implications of theories which may possess additional vector and scalar background fields with arbitrary couplings and ranges in addition to the standard tensor gravitational field.[19]

The general static effective potential between two particles (labelled 1 and 2) is then a straightforward extension of (2.63), viz.

$$V(r) = -\left(\frac{Gm_1m_2}{r} - \frac{g_1^V g_2^V}{4\pi r} e^{-r/\lambda_V} + \frac{g_1^S g_2^S}{4\pi r} e^{-r/\lambda_S} \right), \qquad (2.65)$$

where we have included a single vector and scalar field with couplings and ranges g^V, λ_V and g^S, λ_S respectively. We could readily add further independent vectors and scalars to $V(r)$ if desired.

If the two particles are in motion with 4-velocities u_1 and u_2, with corresponding relativistic factors γ_1 and γ_2, then this potential is modified:

$$V(r) = -\frac{1}{\gamma_1\gamma_2} \left(\frac{Gm_1m_2}{r} \left(2\,(u_1.u_2)^2 - 1\right) - \frac{g_1^V g_2^V}{4\pi r} u_1.u_2\, e^{-r/\lambda_V} + \frac{g_1^S g_2^S}{4\pi r} e^{-r/\lambda_S} \right). \qquad (2.66)$$

More simply, if we consider one particle at rest, so that $u_1.u_2 = \gamma$ (where γ refers to the relative velocity), we can write simply

---

[19]There are a great many proposals for modified gravity theories, from the original Brans–Dicke scalar-tensor theory [79] to the generalised Horndeski models [80] and further extensions involving vector fields such as the TeVeS [81] and scalar-tensor-vector [82] models. For a recent review, see e.g. [83]. As explained above, since our interest centres on antimatter experiments, we are concerned here with theories with new vector fields coupling directly to a fermion current.

$$V(r) = -\left(\frac{1}{\gamma}\left(2\gamma^2 - 1\right)\frac{Gm_1m_2}{r} - \frac{g_1^V g_2^V}{4\pi r}e^{-r/\lambda_V} + \frac{1}{\gamma}\frac{g_1^S g_2^S}{4\pi r}e^{-r/\lambda_S} \right). \quad (2.67)$$

These forms would also accommodate the $U(1)_{B-L}$ interaction discussed above.

Consider now massless gravivectors and graviscalars with infinite-range potentials and couplings of gravitational strength. In the scenarios envisaged in the extended supergravity theories above [76, 77], the couplings of the tensor, vector and scalar interactions would balance leaving a net zero force between pairs of matter particles.

More realistically [84] (see also the extended development of these ideas by Goldman, Hughes and Nieto [85–87]), we could consider a scenario where the gravivector and graviscalar couplings are equal and these interactions cancel. This would occur generically if the vector and scalar arose from dimensional reduction of a five-dimensional theory, as occurs in some supergravity and BSM models. In this case, the gravitational force between matter particles would obey the usual Einstein tensor gravity, whereas that between matter and antimatter would experience a stronger *attraction* due to the addition of the gravivector and graviscalar forces. This possibility has been extensively promoted to motivate independent free-fall experiments on antimatter.

To assess this scenario, we first have to recognise that we cannot engineer an exact cancellation of the gravivector and graviscalar interactions for both the static and moving cases (2.65) and (2.67) due to the different velocity-dependences of the potential for vectors and scalars. Even in this scenario, therefore, the size of the gravivector and graviscalar interactions would still be constrained to some extent by existing equivalence principle tests on matter. This would limit the possible deviation from WEPff in antihydrogen free-fall experiments.

As a quick estimate, equivalence principle tests using lunar laser ranging of the Earth-Moon system in the Sun's gravitational field establish the validity of WEPff to $1 : 10^{13}$ [88]. Since the relative velocity of the Earth and Moon is approx. 1kms^{-1}, corresponding to $\gamma - 1 \sim 10^{-10}$, this would constrain the size of the vector and scalar couplings in (2.65), (2.67) to give $\Delta g/g \lesssim 10^{-3}$, where Δg is the difference in the gravitational acceleration of matter (g) and antimatter.

Secondly, there are several possibilities for the scalar couplings, not all of which match those allowed for a vector [89]. For example, while the absence of anomalies only allows a vector coupling to the combination $B - L$, a scalar could couple to B or L independently. Alternatively, a graviscalar could couple to the full trace of the energy-momentum tensor $T^\mu{}_\mu$ rather than simply the mass as shown in the supergravity models above. More generally, a fundamental scalar would couple to C (and CPT) even operators whereas the vector coupling, as in (2.62), is C (and CPT) odd.

A further serious difficulty with this scenario is that even if a cancellation could be achieved at the level of fundamental interactions, the most stringent EP tests involve bound states, ranging from nucleons (as bound states of quarks and gluons) and atoms to macroscopic bodies. So while a gravivector may couple directly to the masses of the constituent elementary particles, the corresponding tensor graviton and graviscalar would also couple to the full energy-momentum of the bound state.

Moreover, through loop corrections to the energy states (e.g. the Lamb shift [90]) or through the parton distribution functions describing the nucleon structure, these bound states are already sensitive to the existence of antimatter. These corrections have been analysed in some detail in [89–91], where constraints as low as $\Delta g/g \lesssim 10^{-9}$ are quoted. For example, by considering the motion of electrons in the atom, an improved bound from the differing velocity dependence of scalar and vector interactions of $\Delta g/g \lesssim 10^{-7}$ is quoted in [90].

While there is clearly a significant level of model-dependence in all these analyses, the overall picture is clear that the high precision of existing EP tests places severe constraints on scenarios which attempt to hide new gravitational strength vector and scalar forces so that they manifest themselves only in dedicated antimatter experiments. This perhaps emphasises the importance of aiming at high precision in gravity experiments on antimatter as well as in spectroscopy.

The impact of new 'fifth force' vector and scalar interactions would not be limited to WEPff tests. For example, a long-range $U(1)_{B-L}$ vector interaction would produce an analogue of the Stark effect on the spectrum of hydrogen, and oppositely on antihydrogen. Moreover, WEPc tests would in principle also be sensitive to new long-range vector forces, which would modify the Schwarzschild metric around the Earth to Reissner–Nordström type. Spectroscopic measurements analogous to those measuring a gravitational redshift would then exhibit frequency shifts proportional to the coefficient g_{00} of this new metric. These effects will be considered in the following sections.

2.7 Matter-Antimatter Asymmetry and CPT

One of the most important outstanding issues in cosmology is to understand the origin of matter-antimatter asymmetry, i.e. why the observable universe is composed overwhelmingly of matter with only a negligible antimatter component. The issue is quantified by the observed value of the key cosmological parameter $\eta = n_B/n_\gamma$, the baryon-to-photon ratio. Here, n_B and n_γ (essentially the entropy) are the number densities of baryons and photons in the present universe. The observed value [92, 93] is $\eta \simeq 6.1 \times 10^{-10}$.

Explaining the matter-antimatter asymmetry is frequently cited as a motivation for experimental searches for CPT violation, specifically with antihydrogen. Here, we give a brief appraisal of the possible relevance of CPT violation in explaining the observed asymmetry.

The theoretical requirements for a mechanism to generate matter-antimatter asymmetry were set out by Sakharov [94] in the well-known conditions for baryogenesis (or leptogenesis[20]), viz. the theory must exhibit:

---

[20]A lepton asymmetry generated at high temperatures in the early universe can be converted to the present-day baryon asymmetry through non-perturbative "sphaleron" [95] interactions at the electroweak scale. While violating both B and L separately, these processes still conserve $B - L$

1. B (or L) violation;
2. C and CP violation;
3. processes out of thermal equilibrium.

At zero temperature, the standard model is B and L conserving (up to negligible instanton effects), but for temperatures at or above the electroweak scale, which arise in the early universe, non-perturbative sphaleron-induced processes allow substantial B and L violation [96]. This is the origin of the "electroweak baryogenesis" mechanism, in which baryogenesis would take place during a first-order electroweak phase transition. However, the nature of the electroweak phase transition (which is a smooth crossover in the standard model) and the magnitude of CP violation in the CKM matrix mean that this mechanism cannot reproduce the observed value of the baryon-to-photon ratio within the standard model.

In beyond-the-standard-model (BSM) theories, on the other hand, B or L violation is readily achieved, e.g. through the non-equilibrium decay of heavy gauge bosons or Higgs fields in grand unified theories (GUTs) or the decay of additional R-handed neutrinos. Both of these are theoretivcally well-motivated, models with sterile neutrinos with heavy Majorana masses providing a natural 'see-saw' mechanism for generating the light neutrino masses. BSM modifications can also render the electroweak phase transition first-order, realising the electroweak baryogenesis mechanism. There are many related BSM models of baryogenesis or leptogenesis which, for suitably chosen values of the parameters characterising the theory, can give rise to the observed matter-antimatter asymmetry within a standard cosmological framework. (See [97–99] for a selection of reviews.) Indeed, already in 2009, the review [98] quoted more than 40 viable proposals for generating the matter-antimatter asymmetry.

What then is the relevance of CPT violation for this discussion? The Sakharov conditions were formulated under the *assumption* of exact CPT symmetry. Allowing for CPT violation means that it is possible to establish a matter-antimatter asymmetry *in thermal equilibrium* [100]. That is, the third condition can be replaced by

3. CPT violation.

Note, however, that the first two conditions are still required—even with CPT violation, we still need a BSM theory exhibiting B or L violating interactions as well as C and CP violation.

To illustrate what such a model of leptogenesis could look like, consider the following term for the light neutrinos (restricting initially to one flavour for simplicity) in the minimal Lorentz and CPT violating SME discussed in Sect. 2.2:

$$\mathcal{L}_a = a^{(3)}_\mu \, \overline{\nu_L} \gamma^\mu \nu_L \, . \tag{2.68}$$

The dimension-three operator $\overline{\nu_L} \gamma^\mu \nu_L$ is C violating, CP odd, and CPT odd. It is the conserved current for (neutrino) lepton number, so the space integral of its

---

symmetry. Many explanations of the observed baryon-to-photon ratio therefore focus on leptogenesis as the fundamental mechanism.

time component, $\int d^4x\,\overline{\nu_L}\gamma^0\nu_L$, is the corresponding charge, i.e. lepton number. The coupling $a_0^{(3)}$ therefore acts as a chemical potential. This can also be seen directly from the discussion in Sect. 2.3, where we saw how an interaction of the form $\mathcal{L}_a$ would modify the dispersion relation differently for neutrinos and antineutrinos. At non-zero temperature, this modifies the corresponding particle distributions in exactly the way characteristic of a chemical potential in statistical mechanics.

Now, in the high temperature environment of the early Universe, and providing there exist lepton number violating reactions to maintain thermal equilibrium, this effective chemical potential $\mu = a_0^{(3)}$ implies different equilibrium number densities for neutrinos and antineutrinos,[21] resulting in a net lepton number density $n_\ell \sim \mu T^2$. As the universe cools, the rate of these interactions falls until at some decoupling temperature T_D they fall below the rate of expansion, given by the Hubble parameter [101]. In most scenarios, we expect T_D in the range $100\,\text{GeV} \lesssim T_D \lesssim 10^{12}\,\text{GeV}$, its value determined by the BSM dynamics. At this point, thermal equilibrium is no longer maintained and the lepton number density freezes out at the value $n_\ell \sim \mu T_D^2$. Given that the photon density varies with temperature as $n_\gamma \sim T^3$, the resulting lepton-to-photon ratio is then frozen at the value $\eta_\ell \equiv n_\ell/n_\gamma \sim a_0^{(3)}/T_D$. Provided that T_D is above the electroweak scale, this lepton number asymmetry may be subsequently converted to a baryon number asymmetry by sphaleron interactions, yielding a final baryon-to-photon ratio η diluted by a factor of around 10^2 relative to η_ℓ.

To convert this into a realistic model, we need to remember that, as discussed in Sect. 2.3, in the case of a single fermion flavour we can make a field redefinition in the kinetic term in the Lagrangian to remove the interaction $\mathcal{L}_a$. We therefore need to extend $\mathcal{L}_a$ to three flavours of neutrinos, viz. $\mathcal{L}_a \to (a_\mu^{(3)})_{ij}\overline{\nu_L}^i\gamma^\mu\nu_L^j$, together with flavour mixing.

The mechanism described above is essentially that already studied in detail as "spontaneous leptogenesis", where the role of the coupling a_μ is played by a time-dependent VEV of a scalar field, with the equivalence $a_\mu \leftrightarrow \partial_\mu\phi$, or "gravitational leptogenesis", where $a_\mu \leftrightarrow \partial_\mu R$ and the coupling is replaced by the time derivative of the Ricci scalar in an expanding universe (as described briefly in Sect. 2.5).

The same mechanism could also be employed with flavour-mixed quarks, generating a baryon asymmetry directly. In this case, we could envisage a lower decoupling

---

[21] Explicitly, the net lepton number is found from the statistical distributions (neglecting neutrino masses and curvature effects) as

$$
\begin{aligned}
n_\ell = n_\nu - n_{\bar{\nu}} &= g_\nu \int \frac{d^3\mathbf{p}}{(2\pi)^3}\left[\frac{1}{e^{(E-\mu)/T}+1} - \frac{1}{e^{(E+\mu)/T}+1}\right]\\
&= \frac{g_\nu}{2\pi^2}\int_0^\infty dE\,E^2\left[\frac{1}{e^{(E-\mu)/T}+1} - \frac{1}{e^{(E+\mu)/T}+1}\right]\\
&\simeq \frac{g_\nu}{2\pi^2}\frac{2\mu}{T}T^3\int_0^\infty dx\,x^2\frac{e^x}{(e^x+1)^2} = \frac{1}{3}\mu T^2\,.
\end{aligned}
$$

The photon number density is $n_\gamma = \frac{2\zeta(3)}{\pi^2}T^3$. For comparison, the entropy density, which is alternatively used to normalise the baryon asymmetry, is $s = \frac{2\pi^2}{45}g_{*s}T^3$, where g_{*s} is the effective number of light degrees of freedom at the scale T.

temperature, but an absolute requirement is that the baryon asymmetry must be established before the onset of big-bang nucleosynthesis at $T = 10\,\text{MeV}$, the successful theory of which provides a stringent constraint on the value of η.

At this point, we can check whether this mechanism could be realistic, given the experimental constraints on the relevant minimal SME parameters. Of course, this makes the assumption that these couplings remain constant over the evolution of the Universe. This is non-trivial, since if they are regarded as VEVs of some time-dependent fields, their values would evolve and could be markedly higher at the time of lepto(baryo)genesis than their current values. From the SME data tables [44] (Table D26), bounds on the relevant neutrino coefficients of $(a_0^{(3)})_{ij} \lesssim 10^{-20}\,\text{GeV}$ are quoted. With $T_D > 100\,\text{GeV}$, the resulting prediction for $\eta_\ell \sim a_0^{(3)}/T_D$ would be many orders of magnitude too small. Constraints on $(a_0^{(3)})_{ij}$ in the quark sector would similarly exclude the direct baryogenesis scenario under these assumptions.

However, in [102], where this type of CPT violating model was first considered, the potential of higher-dimension operators in the SME effective Lagrangian to produce the observed value of the asymmetry was discussed. For example, we could consider interactions of the form $\mathcal{L}_a^{(5)} = -a_{\mu\rho\sigma}^{(5)}\bar{\psi}\gamma^\mu\partial^\rho\partial^\sigma\psi$, where here we can simply take ψ to be the electron field. This would be the leading-order electron coupling of this type, since the dimension 3 coupling $a_\mu^{(3)}$ can be removed by a field redefinition and is not physical. (This interaction is described in the context of $\overline{\text{H}}$ spectroscopy in Sect. 3.2.2 below.) Because of the extra derivatives in the operator, the same analysis would predict a corresponding lepton asymmetry of order $|a_{0\rho\sigma}^{(5)}|\,T_D$.

Now, in the spirit of effective actions, where higher order couplings should be suppressed by powers of the fundamental UV scale M, this may be expected to give a smaller contribution than $a_0^{(3)}/T_D$ by a power of T_D^2/M^2. However, if we set this aside, we could simply ask what the current experimental bounds on the non-minimal couplings would allow. In order to extract phenomenology from this operator, we can either make a non-relativistic expansion, which leads amongst others to the coupling a_{200}^{NR} which enters the transition frequency for the $1S$–$2S$ transition in H and $\overline{\text{H}}$ (see Sect. 3.2.2), or an ultra-relativistic expansion [103] in which the relevant parameter is $\mathring{a}^{\text{UR}(5)}$. The ultra-relativistic regime is appropriate for an analysis of the astrophysical effects of modified dispersion relations, which would allow reactions such as $\gamma \to e^+e^-$ which are ruled out by the observation of multi-TeV gamma rays [103, 104]. The couplings a_{200}^{NR} and $\mathring{a}^{\text{UR}(5)}$ are closely related, but not identical, and very different bounds for the electron are quoted in the the SME data tables [44] (Table D7). Neither corresponds exactly to the combination of $a_{\mu\rho\sigma}^{(5)}$ couplings determining the lepton asymmetry in the model above. Nevertheless, it is clear that the measurement of a non-vanishing a_{200}^{NR} coefficient in the well-understood setting of $1S$–$2S$ $\overline{\text{H}}$ spectroscopy would provide a strong impetus to models of leptogenesis and baryogenesis based on CPT violation.

Finally, note that here we have only considered a relatively simple model for illustration. Other possibilities combining CPT violation, also involving other SME couplings, and more exotic BSM physics can readily be constructed and have been actively studied (see e.g. [105] for examples and a review) and may be added to the extensive list of proposals for a theory of baryogenesis.

To summarise this discussion, we have seen several general points to keep in mind while considering the relevance of CPT violation to leptogenesis and baryogenesis:

(i) There are already a large number of realistic proposed mechanisms of baryo-genesis *without* CPT violation, but they all need some BSM physics.

(ii) With CPT violation, a baryon or lepton asymmetry may be generated *in thermal equilibrium*, in contrast to CPT conserving theories, via a mechanism similar to spontaneous or gravitational baryo(lepto)genesis. However, these models still require new BSM physics with as yet undiscovered particles.

(iii) Existing constraints on SME parameters appear to rule out models restricted to the minimal SME. Non-minimal SME couplings are in general less constrained, however, and the different dependence on the decoupling temperature in models involving them offers the potential to achieve the required asymmetry, provided the couplings are not too suppressed by the high-energy scale underlying the SME effective Lagrangian.

(iv) In the minimal SME, the parameters which are probed in $\overline{\text{H}}$ spectroscopy (notably the space components $b_3^{e,p}$) are different from those (e.g. the time components $a_0^{\nu,q}$) which enter the simplest leptogenesis or baryogenesis mod-els. At higher order, the non-minimal couplings accessible to spectroscopy are very closely related, though not identical, to those involved in the sim-plest models of baryo(lepto)genesis. Any measurement of CPT violation in $\overline{\text{H}}$ spectroscopy would therefore have a significant impact on the development of theories seeking to explain the matter-antimatter asymmetry of the universe.

Chapter 3
Antihydrogen

So far, only the ALPHA collaboration has performed experiments to measure some of the properties of antihydrogen relevant for testing the fundamental physics described above. Accordingly, we will use their data as exemplars in our discussion. As summarised in Chap. 1, the ALPHA results to date comprise: a limit on the charge neutrality of the anti-atom [30, 31]; demonstration of a method to investigate the gravitational interaction of antihydrogen [34] (a study accompanied by detailed investigations of the trajectories of antihydrogen atoms held in a magnetic minimum neutral atom trap [32, 33]); detection and spectroscopy of antihydrogen ground state hyperfine transitions [35, 36], the two-photon Doppler-free $1S$–$2S$ transition [1, 2], and the observation of the Lyman-α ($1S$–$2P$) transition [28] and a determination of the Lamb shift [106] in the anti-atom.

These investigations fall into two broad categories. As fundamental physics tests, the spectroscopy measurements and charge limits are primarily sensitive to potential Lorentz and CPT violations, though in principle could also be affected by a new $U(1)_{B-L}$ force. The gravitational studies are tests of WEPff, including 'fifth forces'. Clearly these are free-fall experiments, but it should be kept in mind that even the charge investigations also involve assumptions regarding the form and evolution of the trajectories of the trapped anti-atom which could be influenced by a WEPff violation.

3.1 Charge Neutrality of Antihydrogen

One of the most direct tests of fundamental principles with antihydrogen, and one of the first to be confirmed experimentally [31], is its electric charge neutrality. Evidently, the net electric charge of the anti-atom is expected to be zero, interpreted as the sum of the charges of the antiproton (-1) and positron ($+1$). The question we address here is what fundamental principles are tested by an experimental measurement of a putative antihydrogen charge, here denoted as Q.

© The Author(s), under exclusive license to Springer Nature Switzerland AG 2020
M. Charlton et al., *Antihydrogen and Fundamental Physics*,
SpringerBriefs in Physics, https://doi.org/10.1007/978-3-030-51713-7_3

3.1.1   Antihydrogen Charge Measurement

ALPHA has performed two studies that have been used to set a limit on Q via an analysis of the behaviour of the anti-atom in the presence of applied electric (E) and magnetic (B) fields, were it to have a charge. The first [30] consisted of a retrospective analysis of an experiment in which trapped antihydrogen atoms were released from the neutral atom trap via magnetic field reduction, in the presence of electric fields (biased in different trials either axially to the left or the right of the magnetic trap centre, and denoted as E_L and E_R respectively) whose function was to aid in distinguishing antihydrogen annihilations from those of any antiprotons remaining in the trap region. A search was made for a possible electric field-induced shift in the measured axial antihydrogen annihilation distributions, $\langle z \rangle_\Delta$, caused by a non-zero $Q \propto \langle z \rangle_\Delta / (E_R - E_L)$. From a measured mean axial deflection of 4.1 ± 3.4 mm from the centre of the atom trap a limit was found as $Q = (-1.3 \pm 1.1 \pm 0.4) \times 10^{-8} e$ including statistics and systematics, with a 1σ confidence.

In their second experiment [31] ALPHA used a stochastic heating method in which the application of random time-varying electric fields would result in stochastic energy kicks to a charged anti-atom such that it would eventually leave its shallow trapping well. In particular, an antihydrogen atom with charge Q would gain (from N kicks of voltage change $\Delta\phi$) a kinetic energy of $|Q|e\Delta\phi\sqrt{N}$ such that it will leave a well of depth E_{well} if $|Q| \gtrsim E_{well}/e\Delta\phi\sqrt{N}$. The measured parameter was the survival probability of the anti-atoms in the trap, when compared to null trials when no stochastic field was applied, and the result was that the charge was bounded as $|Q| < 0.71$ ppb (1σ): a 20-fold improvement on the previous study [30].

In both of these experiments simulations of antihydrogen trajectories were used extensively in the derivation of the final results for Q, and in studies of the pertinent uncertainties—particularly those due to systematic effects. Specifically, it was assumed that the antihydrogen motion is such that its magnetic moment adiabatically follows the (spatially varying) magnetic field of the neutral trap such that its motion can be investigated using classical mechanics. The combined equation of motion for the centre of mass position $\mathbf{r}$ of antihydrogen, with an inertial mass m_i and charge Q (in units of the fundamental charge e) and so-called gravitational mass m_g (see Sect. 3.3.1), in spatially and temporally varying electric and magnetic fields ($\mathbf{E}(\mathbf{r}, t)$ and $\mathbf{B}(\mathbf{r}, t)$ respectively) is,

$$m_i \frac{d^2\mathbf{r}}{dt^2} = \nabla(\boldsymbol{\mu} \cdot \mathbf{B}(\mathbf{r}, t)) + Qe\left[\mathbf{E}(\mathbf{r}, t) + \frac{d\mathbf{r}}{dt} \times \mathbf{B}(\mathbf{r}, t)\right] - m_g g\hat{\mathbf{y}}, \qquad (3.1)$$

where $\boldsymbol{\mu}$ is the magnetic moment and g is the local acceleration due to gravity which is assumed to act in the y-direction. Here the familiar, essentially Newtonian, notion that a possible WEP anomaly can expressed via a "gravitational" mass m_g distinct from the inertial mass has been applied, with ALPHA defining the ratio $F = m_g/m_i$ and setting limits on this parameter from an analysis of their experiment, as will be described below.

Ahmadi et al. [31] have also described how this value for Q can be used in conjunction with the limit on the so-called antiproton charge anomaly [107] (defined in obvious notation as $(|q_{\bar{p}}| - e)/e$) to reduce the corresponding quantity for the positron by a factor of 25 to 1 ppb. (1σ). The assumption in the analysis is that the value of m_i used is that of hydrogen.

Such charge anomaly limits are usually quoted as tests of CPT (see e.g. [93]), since, as described in Sect. 2.1, charge differences equal to zero between particles and antiparticles (and hence $Q = 0$) are strict requirements of the CPT theorem. We comment critically on this below. We also draw attention here to the work of Hughes and Deutch [108] who used measurements of cyclotron frequencies and spectroscopic data to derive limits on the electric charges of positrons and antiprotons. The direct limit set on Q described here could also be used to augment and update the jigsaw of data they present.

3.1.2 Theoretical Principles

Of course, Lorentz and CPT symmetry does imply that the charges of $\overline{H}$ and H should be the same. However, charge neutrality of antihydrogen is not really a test of CPT in the sense that even in the Lorentz and CPT violating effective theories described in Chap. 2, the charges of the proton and antiproton, and the electron and positron, are still required to be equal and opposite. In fact, this is a much deeper property of relativistic QFT which ultimately is necessary for *causality*.

As explained in Sect. 2.1, the existence of antiparticles with precisely the same mass and opposite charge is a requirement of causality, necessary to ensure the vanishing of the relevant correlators for spacelike separation (microcausality). It is therefore almost impossible to see how any difference in magnitude of the charge of the electron and positron could be compatible with our present understanding of QFT.

Even if we momentarily set these structural issues aside and continue to entertain the idea that the electron and positron charges could be different, further difficulties proliferate. Given the existence of the coupling of the photon to an $e^+ e^-$ pair, either the photon would be charged or, in contradiction to a huge body of experimental evidence, electric charge conservation would be violated. In the former case, an accelerating electron would lose charge by synchrotron radiation, leading to further paradoxes. Indeed, experimental limits on the charge of the photon are extremely strong, the PDG quoting a bound of $10^{-35}e$ [93].

A second issue, which would apply equally to hydrogen, is then the equality of the magnitude of the charges of the antiproton and positron (or equivalently, the proton and electron). In the standard model, this is again ensured by a deep principle, in this case *unitarity*. As discussed in Sect. 2.6, the absence of anomalies in gauged currents imposes a set of constraints on the charges of the fermions which appear as internal lines in the 3-current triangle diagrams. This imposes a precise balance between the quark and lepton charges. If this is broken, we would lose conservation of a gauged current, which in turn would lead to a loss of unitarity in the QFT.

We see, therefore, that the experimental measurement of charge neutrality of antihydrogen should be viewed as a test of *causality* and *unitarity* in the standard model QFT. While CPT invariance does imply the equality of the charge of hydrogen and antihydrogen, charge neutrality would still be required even in a CPT-violating QFT.

The only remaining loophole would seem to be our assumption that the measured charge of the bound states is given precisely by the sum of the charges of their constituents, either the antiproton and positron for the antihydrogen atom or the valence quarks $\bar{u}\bar{u}d$ for the antiproton. This is referred to as the "assumption of charge superposition" in [31]. It could conceivably be possible to imagine some sort of charge screening mechanism which could invalidate this for a particular experimental measurement. However, such a screening effect would have had to avoid detection everywhere else in the particle physics of hadrons, or indeed in hydrogen, and seems extremely unlikely to be a factor in interpreting the antihydrogen charge neutrality experiment. Fortunately, these experiments have indeed validated the expected null result to high precision.

3.2 Antihydrogen $1S$–$2S$, $1S$–$2P$ and $1S$ Hyperfine Spectroscopy

The advent of precision antihydrogen spectroscopy has opened a new window to test fundamental principles such as Lorentz and CPT invariance, to extend experimental tests of GR to antimatter systems, and to search for new long-range forces.

The gold standard spectroscopic measurement in this field is the two-photon $1S$–$2S$ transition in hydrogen, for which the transition frequency is known to a few parts in 10^{15} [109], many orders of magnitude more precise than state of the art theoretical calculations in QED. A direct comparison of the antihydrogen and hydrogen spectra therefore provides a more precise test of QED and CPT invariance than follows from theory and hydrogen spectroscopy alone, further motivating these antimatter experiments.

3.2.1 Antihydrogen Spectroscopy

Here, we give a brief overview of the experimental arrangement for spectroscopy in the ALPHA apparatus. Details, including a level diagram (Fig. 3.1) with state labels, of the transitions involved in spectroscopy are given in Sect. 3.2.2 below.

Historically, only microwave radiation could be coupled into the ALPHA apparatus, giving access to spin-flip transitions between ground-state hyperfine levels by positron spin resonance (PSR) [35]. These transitions result in a reversal of the magnetic moment of the antihydrogen atom which will then no longer be trapped, leading to an annihilation signal registered by the silicon vertex detector. In addition to being the first ever interrogation of the internal structure of an anti-atom, this experiment is also of great significance because it shows that a spectroscopic signal can be produced from only one trapped anti-atom. The experiment also gave rise to the appearance and disappearance detection protocols described in Chap. 1, which have been used in various combinations in all subsequent spectroscopic measurements. The ALPHA-2 upgrade includes a waveguide which enables efficient delivery of microwave radiation directly into the trap structure, and thereby higher transition rates. The microwave radiation is also used to drive the electron cyclotron resonance in trapped plasmas to determine the magnitude of the central magnetic field (approx. 1 T). The field can be determined with a precision of 1 ppm and an accuracy of 4 ppm. When the microwaves are tuned to the relevant PSR frequency (just below 30 GHz at the B-field used in experiments), any unwanted population of hyperfine level c states can be efficiently removed, allowing spectroscopy with spin polarised samples.

Ground-state hyperfine spectroscopy is performed by first trapping ground-state antihydrogen atoms. During antihydrogen synthesis there is no control of the internal state of the anti-atom, and the trapped population typically contains equal amounts of anti-atoms in the weak-field seeking $1S_c$ and $1S_d$ states. Spectroscopy proceeds by injecting microwave radiation and stepping the frequency of the microwaves in small intervals over a frequency range (starting below resonance) that includes the onset of the $1S_c$–$1S_b$ transition which is effectively set by the magnitude of the central magnetic trap B-field. Once the frequency has been stepped over a range that covers all main spectral features, the frequency is then adjusted up by the hyperfine splitting (approx 1.42 GHz) and the stepping continues to cover the $1S_d$–$1S_a$ transition in a similar fashion. Antihydrogen is detected in appearance mode during the frequency sweep, producing a spectrum which shows a rapid onset of resonance for both transitions. The onset frequency difference is used to determine the hyperfine splitting, leading to an uncertainty of four parts in 10^4. A refined measurement, making full use of the electron cyclotron resonance technique to determine the B-field, is possible.

The $1S$–$2S$ transition has long been the gold-standard in laser spectroscopy of atomic hydrogen due to the long lifetime (about $1/8$ s) of the $2S$ state, and the cancellation of the Doppler effect (to first order) when a transition from the $1S$ state is induced by two counter-propagating photons. The flip-side is that, in addition to a narrow bandwidth, substantial laser power is required to drive the dipole-forbidden transition from the $1S$ ground state with experimentally relevant rates. While narrow band sources can nowadays straightforwardly be created from diode lasers, it is technically challenging to achieve sufficient power at the 243 nm (UV) wavelength of the two-photon transition. In the ALPHA experiment the power from commercial laser sources is resonantly enhanced in an optical cavity surrounding the trapping region in a near-axial orientation. The enhancement cavity which is operated in the cryogenic region near 4 K and in ultra-high vacuum, yields about 1 W–2 W of circulating laser power with a 200 μm waist which is sufficient to cause photoionisation from the $2S$ state with a third photon. The ionised antihydrogen atom is no longer confined and thus, a signal on the silicon vertex detector is produced when the laser is on resonance. The laser system is referenced to atomic time via a GPS-scheduled quartz oscillator, with additional referencing from a locally operated Cs atomic clock, to provide the frequency accuracy of 8×10^{-13}. The transition was observed [1] and later characterised with an uncertainty of 2 parts in a trillion [2].

While the $1S$–$2P$ transition in hydrogen and antihydrogen is arguably the simplest transition in any atomic system, the 121.6 nm wavelength (the famous Lyman-alpha line) poses a formidable challenge for laser spectroscopy since there are no simple laser systems in this region of the spectrum. The short (about 1.6 ns) lifetime of the $2P$ state leads to a broad natural linewidth even before any further inhomogeneous broadening such as the Doppler effect. The Lyman-alpha line is nevertheless of great significance e.g. in astronomy and cosmology. For precision experiments with hydrogen and antihydrogen the line is of relevance because scattering red-detuned photons on the transition leads to Doppler cooling which in turn reduces line broadening. In order to achieve both a reliable and sufficiently intense source for excitation of the line, ALPHA has constructed a pulsed laser system based upon frequency tripling in Kr/Ar gas. The fundamental wavelength is produced from a pulsed solid-state (Ti:sapphire) laser which is seeded from a narrow-band diode laser. The current system produces up to 0.8 nJ inside the trapping region per 12 ns pulse at 121.6 nm, with a 65 MHz linewidth. Due to positron spin-mixing (see Sect. 3.2.2), the excited atom may decay quickly to an untrapped state and a signal is produced on the silicon vertex detector. An added advantage of the pulsed laser system is that the appearance signal can be analysed in coincidence with the arrival of the pulse, leading to a measurement of the time-of-flight from the trap to the electrode wall, from which the velocity distribution of the trapped atoms can be reconstructed. The observation of the Lyman-alpha transition [28], which was performed with 500 accumulated antihydrogen atoms, paves the way for antihydrogen fine structure spectroscopy and laser cooling, while more recently the ALPHA collaboration has presented a detailed investigation of the fine structure of the $1S$–$2P$ transitions at a precision of 16 parts per billion, allowing a first determination of the Lamb shift for antihydrogen [106].

3.2.2 Lorentz and **CPT** Violation

In this section, we analyse the possibility of observing Lorentz or **CPT** violation in antihydrogen spectroscopy, focusing on the $1S$–$2S$, $1S$–$2P$ and the $1S$ hyperfine transitions. We follow the systematic approach of parametrising potential Lorentz/CPT violating effects in terms of the couplings in the SME effective action. Our discussion here is not new,[1] the aim being simply to provide an accessible presentation of the key principles involved.

---

[1] In this spirit, we give here only a limited set of references to enable the reader to follow the quoted results. For a more complete survey of the extensive SME literature and its applications to antimatter and spectroscopy, see for example the review and compendium of limits on the couplings in [44].

The starting point is to approximate the full SME effective action by the corresponding non-relativistic Hamiltonian appropriate for low-energy atomic physics. This has been carried out by several authors [110–114], including not just the coefficients of the renormalisable SME operators shown in (2.19) but also those from the full expansion of higher derivative operators in the extended effective action [103, 115, 116].

The non-relativistic Hamiltonian is obtained from the SME Lagrangian (2.19) in the standard way as an expansion in powers of p_i/m_e, where p_i is the electron/positron momentum, the novelty being simply that the familiar Dirac and QED action is extended to include the additional Lorentz and CPT-violating couplings. The leading terms in this expansion contribute at $O(1)$ in the fine structure constant to the atomic energy levels, while those of $O(p_i p_j/m_e^2)$ give corrections of $O(\alpha^2)$. As we shall see, the SME corrections to the $1S$–$2S$ transition frequency measured by ALPHA only occur at $O(\alpha^2)$.

Following [111, 112, 114], we can write the relevant terms in the SME modified Hamiltonian appropriate for describing the antihydrogen atom as:

$$H_{\text{SME}} = \sum_{\omega=e,p} \left[A^\omega + 2B_k^\omega S^k + \left(E_{ij}^\omega + 2F_{ijk}^\omega S^k \right) \frac{p_i p_j}{m_e^2} + \cdots \right], \qquad (3.2)$$

where $S^k = \tfrac{1}{2}\sigma^k$ is the spin operator. In terms of the SME Lagrangian couplings,

$$
\begin{aligned}
A^e &= -a_0^e - m_e c_{00}^e + \cdots \\
B_k^e &= -b_k^e - m_e d_{k0}^e - \tfrac{1}{2}\epsilon_{kij} H_{ij}^e \\
E_{ij}^e &= -m_e \left(c_{ij}^e + \tfrac{1}{2}c_{00}^e \delta_{ij} \right) \\
F_{ijk}^e &= -\tilde{d}_i^e \delta_{jk} + \tfrac{1}{2} \left(\delta_{ij}b_k^e - \delta_{ik}b_j^e \right) + \cdots
\end{aligned}
\qquad (3.3)
$$

with $\tilde{d}_i^e = m_e \left(d_{0i}^e + \tfrac{1}{2}d_{i0}^e \right) - \tfrac{1}{4}\epsilon_{ijk} H_{jk}^e$. Analogous expressions hold for the antiproton, but with an overall extra factor of $\epsilon = m_e^2/m_p^2$ in E_{ij}^p and F_{ijk}^p to compensate for the use of the electron mass in the momentum factor in (3.2). As in (2.19), we omit the SME coefficients e^μ, f^μ and $g^{\lambda\mu\nu}$ essentially for ease of presentation (see also Footnote 6 in Chap. 2). Compared with the corresponding Hamiltonian for hydrogen with electrons and protons, we have simply made the substitutions $a_\mu \to -a_\mu$, $b_\mu \to b_\mu, c_{\mu\nu} \to c_{\mu\nu}, d_{\mu\nu} \to -d_{\mu\nu}, H_{\mu\nu} \to -H_{\mu\nu}$ dictated by the C conjugates of the corresponding operators.

So far, we have considered only the operators of dimension ≤ 4 in the SME Lagrangian, corresponding to a renormalisable theory (the *minimal* SME). However, as discussed in Sect. 2.3, we should rather view the SME as an effective field theory describing the low-energy dynamics of some UV-complete fundamental QFT. The effective theory also includes (non-minimal) operators of dimension >4, with the corresponding dimensional couplings being suppressed by powers of $1/\Lambda$, where Λ is the scale of the fundamental theory, and so provides a systematic expansion for corrections to the leading-order low-energy physics.

To illustrate this, we focus on one particular dimension 5 operator which will play a rôle in describing the $1S$–$2S$ antihydrogen transitions, viz.

$$L_{SME}^{(5)} \sim \int d^4x \; a_{\mu\rho\sigma}^{(5)} \; \bar{\psi}\gamma^\mu D^\rho D^\sigma \psi \; + \; \cdots \, , \tag{3.4}$$

and consider its relevant contribution to the non-relativistic Hamiltonian for antihydrogen,

$$H_{SME}^{(5)} \sim - \sum_{\omega=e,p} \left(a^{(5)\,\omega} + a_{ij}^{(5)\,\omega} \, p_i p_j \; + \; \cdots \right). \tag{3.5}$$

Note that the detailed relation of the coefficients denoted $a^{(5)}$ and $a_{ij}^{(5)}$ here to the components $a_{\mu\rho\sigma}^{(5)}$ in (3.4) is not especially simple. A fully comprehensive and systematic account of these higher-dimensional operators and their contribution to hydrogen and antihydrogen spectroscopy is given in [103, 115]. The terms in (3.5) can then be added to the Hamiltonian (3.2) where they correspond to the substitutions $A \to A - a^{(5)}$ and $E_{ij} \to E_{ij} - m_e^2 a_{ij}^{(5)}$. Comparing with the c_{00} and c_{ij} terms, we see that in the effective field theory expansion this term would be expected to be suppressed by $O(m_e/\Lambda)$. In the usual interpretation of the SME as the low-energy effective theory corresponding to a fundamental theory at the string or Planck scales, this is clearly a tiny factor. Nevertheless, we will carry it forward in the analysis of the antihydrogen transitions below.

To include the contributions from the Lorentz and **CPT** violating operators in the photon sector in (2.19), we need a slightly different approach. First note that these terms in the SME Lagrangian modify the photon propagator through two-point interactions $-2i(k_F)_{\mu\rho\nu\sigma}q^\rho q^\sigma$ and $2(k_{AF})^\rho \epsilon_{\mu\rho\nu\sigma}q^\sigma$ respectively (where q^μ is the momentum of the photon propagator). This modifies the photon mediated interaction between the antiproton and the positron, changing the energy levels of the atom. This can be calculated in the framework of non-relativistic QED [117], the result quoted in [114] being a change in the energy of a given state $|\psi_n\rangle$ given by

$$E_{SME}^\gamma = \frac{\alpha^2}{n^2} \langle \psi_n | m_e (k_F)_{0i0j} \left(\hat{p}_i \hat{p}_j - \delta_{ij} \right) + (k_{AF})_i L_i | \psi_n \rangle \, , \tag{3.6}$$

where L_i is the positron orbital angular momentum and $\hat{p}_i$ is its unit 3-momentum vector. However, this only includes the *spin-independent* contributions to the effective positron-antiproton potential and a complete analysis of the bound state involves many new vertices in the NRQCD framework [118]. In particular, including the photon coupling to the so-called Fermi vertex gives a further k_{AF} dependent contribution of (compare [119]),

$$E_{SME,\,spin}^\gamma = \frac{\alpha^2}{n^2} \langle \psi_n | (k_{AF})_i \left(S_i + \hat{p}_i \hat{p}_j S_j \right) | \psi_n \rangle \, . \tag{3.7}$$

Evidently, these terms can be absorbed into extra k_{AF} dependent pieces in B_k^e and F_{ijk}^e in the Hamiltonian (3.2). Their contribution to the antihydrogen transitions considered below can therefore easily be included, but we do not display them explicitly here since, with hindsight, k_{AF} is known to be bounded far more stringently from astrophysics than the known bounds from atomic physics on the matter terms in (3.2), especially b_k^e.

$1S$ Hyperfine Transitions

To calculate the contributions of these Lorentz and CPT violating couplings to the antihydrogen spectrum and the specific transitions measured by ALPHA, we need first to describe the appropriate hyperfine states, remembering that in ALPHA the anti-atoms are confined in a magnetic trap. Our notation is $\mathbf{F} = \mathbf{I} + \mathbf{J}$, where $\mathbf{I}$ is the antiproton spin and $\mathbf{J} = \mathbf{L} + \mathbf{S}$, where $\mathbf{L}$ and $\mathbf{S}$ are the positron orbital and spin angular momentum respectively. States are labelled by the corresponding quantum numbers as $|n\, \ell\, j\, f\, m_f\rangle$. Restricting initially to $\ell = 0$ states (for which $j = 1/2$), we will use both this 'hyperfine' basis $|f\, m_f\rangle$ and the 'spin' basis $|m_I\, m_s\rangle$ as convenient. Following [36], we also denote the antiproton spin $m_I = \pm 1/2$ by $\Uparrow$, $\Downarrow$ and the positron spin $m_s = \pm 1/2$ by $\uparrow$, $\downarrow$.

The hyperfine interaction coupling the antiproton and positron spins is determined by the Hamiltonian $H_{\text{hyp}} = \mathcal{E}_{HF}\, \mathbf{I}.\mathbf{S}$. An elementary calculation shows that the degeneracy of the $1S$ states is split, with $\Delta E = -(3/4)\mathcal{E}_{HF}$ for the singlet state $|f\, m_f\rangle = |0\,0\rangle$ while $\Delta E = (1/4)\mathcal{E}_{HF}$ for the triplet $|1\, m_f\rangle$ with $m_f = \pm 1, 0$. $\mathcal{E}_{HF}$ is therefore the hyperfine splitting of the $f = 1$ and $f = 0$ states at zero magnetic field. Explicitly,

$$\mathcal{E}_{HF} = \frac{2}{3}\, \mu_0\, g_e\, \mu_B\, g_p\, \mu_N\, |\psi_{n00}(0)|^2\,, \tag{3.8}$$

where $\mu_B = e/2m_e$ is the Bohr magneton, $\mu_N = e/2m_p$, and the g-factors are approximately $g_e = 2.002$ and $g_p = 5.585$. $\psi_{n00}(0)$ is the nS wave function at the origin, and $|\psi_{n00}(0)|^2 = 1/(\pi n^3 a_0^3)$ where $a_0 = 1/(\alpha m_e)$ is the Bohr radius.

In a constant magnetic field $\mathbf{B}$, the Hamiltonian including the Zeeman coupling becomes

$$H_{\text{hyp}} = \mathcal{E}_{HF}\mathbf{I}.\mathbf{S} - g_e\mu_B\, \mathbf{S}.\mathbf{B} + g_p\mu_N\, \mathbf{I}.\mathbf{B}\,. \tag{3.9}$$

The energy eigenstates at non-zero $\mathbf{B} = B\mathbf{e}_z$ are linear combinations of the $m_f = 0$ states, found by diagonalising the Hamiltonian (3.9).

In the spin basis, they are

$$
\begin{aligned}
|d\rangle &= |\Downarrow\, \downarrow\rangle \\
|c\rangle &= \cos\theta_n |\Uparrow\, \downarrow\rangle + \sin\theta_n |\Downarrow\, \uparrow\rangle \\
|b\rangle &= |\Uparrow\, \uparrow\rangle \\
|a\rangle &= -\sin\theta_n |\Uparrow\, \downarrow\rangle + \cos\theta_n |\Downarrow\, \uparrow\rangle\,,
\end{aligned}
\tag{3.10}
$$

where the mixing angle $\theta_n(B)$ is given by $\tan 2\theta_n = \mathcal{B}_0/n^3 B$, with $\mathcal{B}_0 = \mathcal{E}_{HF}/g_e\mu_B$ $(1 + \epsilon_p) \simeq 50.7\,\mathrm{mT}$ [110]. Here, we have defined $\epsilon_p = g_p m_e/g_e m_p$. The corresponding form in the hyperfine basis is given in the footnote.[2]

Clearly, for zero field we have $\cos\theta_n = 1/\sqrt{2}$ and the $|c\rangle$ and $|a\rangle$ states are just the hyperfine $|1\ 0\rangle$ and $|0\ 0\rangle$ states respectively. In ALPHA, the magnetic field is of order $B \sim 1\mathrm{T}$ which is in the high field regime where $\cos\theta_n \simeq 1$, so the states are very well approximated by $|c\rangle \simeq |\Uparrow\ \downarrow\rangle$ and $|a\rangle \simeq |\Downarrow\ \uparrow\rangle$.

The corresponding energy levels are

$$E_{d,b} = \frac{1}{4}\mathcal{E}_{HF} \pm \frac{1}{2}g_e\mu_B B(1 - \epsilon_p)$$

$$E_{c,a} = -\frac{1}{4}\mathcal{E}_{HF} \pm \frac{1}{2}\left[\mathcal{E}_{HF}^2 + \left(g_e\mu_B B(1 + \epsilon_p)\right)^2\right]^{1/2}, \qquad (3.11)$$

reproducing the well-known Breit–Rabi formula [120].[3]

These energy levels are shown for the $1S$ antihydrogen states in the lower panel of Fig. 3.1. In the ALPHA trap, only the two 'low-field seeking' states $|d\rangle$ and $|c\rangle$ are trapped, while the 'high-field seeking' states $|b\rangle$ and $|a\rangle$ escape the trap.

The ground state hyperfine energy levels receive corrections due to the Lorentz and **CPT** violating couplings at $O(1)$ in the fine structure constant. We therefore need only calculate the contributions of the A and B_k terms in (3.2) to these levels. For this, we need the expectation values of the antiproton and positron spin operators. These are most simply evaluated using the spin basis, e.g.

$$\langle c| S_z^e |c\rangle = (\cos\theta_1\langle\Uparrow\ \downarrow| + \sin\theta_1\langle\Downarrow\ \uparrow|)\, S_z^e\,(\cos\theta_1|\Uparrow\ \downarrow\rangle + \sin\theta_1|\Downarrow\ \uparrow\rangle)$$

$$= -\frac{1}{2}\cos 2\theta_1, \qquad (3.12)$$

using $S_z^e|\Uparrow\ \downarrow\rangle = -\frac{1}{2}|\Uparrow\ \downarrow\rangle$, etc. The full set of SME corrections are then found to be

---

[2]For convenience, we also give the explicit form of these states in the hyperfine basis:

$$|d\rangle = |1\ -1\rangle$$
$$|c\rangle = \frac{1}{\sqrt{2}}((\cos\theta_n + \sin\theta_n)\,|1\ 0\rangle - (\cos\theta_n - \sin\theta_n)\,|0\ 0\rangle)$$
$$|b\rangle = |1\ 1\rangle$$
$$|a\rangle = \frac{1}{\sqrt{2}}((\cos\theta_n - \sin\theta_n)\,|1\ 0\rangle + (\cos\theta_n + \sin\theta_n)\,|0\ 0\rangle).$$

[3]Here, we seek the leading order form of the states for calculating expectation values in the SME. We note that in order to work out precise energies and energy differences, such as those required for two-photon spectroscopy, corrections due to the difference in the magnetic moment and the diamagnetic term need to be taken into account.

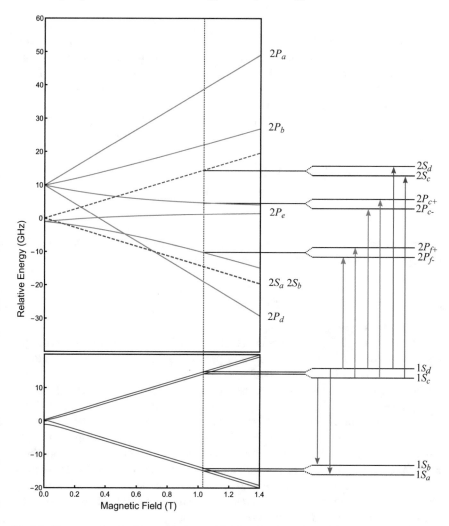

Fig. 3.1 The energy levels of the antihydrogen $1S$, $2S$ and $2P$ states in a uniform magnetic field, with the central magnetic field in the ALPHA apparatus indicated by a vertical dashed line. The origins of the vertical axis in the upper and lower panels are separated by the $1S$–$2S$ energy difference. Transitions discussed in the text are denoted by arrows from the initial state in experiments. Channels to final states in transitions excited during laser spectroscopy are omitted for clarity

$$
\begin{aligned}
E_{\text{SME}}^{d} &= A^e + A^p - \left(B_3^e + B_3^p\right) \\
E_{\text{SME}}^{c} &= A^e + A^p - \cos 2\theta_1 \left(B_3^e - B_3^p\right) \\
E_{\text{SME}}^{b} &= A^e + A^p + \left(B_3^e + B_3^p\right) \\
E_{\text{SME}}^{a} &= A^e + A^p + \cos 2\theta_1 \left(B_3^e - B_3^p\right) \ .
\end{aligned}
\tag{3.13}
$$

At zero magnetic field, $\cos 2\theta_1 \to 0$, and the $|c\rangle$ and $|a\rangle$ states are only shifted by the $A^e + A^p$ term common to all states. The $|c\rangle \to |a\rangle$ transition is therefore insensitive to the Lorentz and CPT violating couplings in the minimal SME and simply measures the hyperfine splitting $\mathcal{E}_{HF}$. In fact, this result holds even in the non-minimal theory [115].

The two transitions measured by ALPHA [36] are $|d\rangle \to |a\rangle$ and $|c\rangle \to |b\rangle$, both from trapped to untrapped states (red arrows in Fig. 3.1). The SME contributions to these transition frequencies are

$$
\begin{aligned}
\Delta E_{d\leftrightarrow a}^{\bar{\text{H}},\,\text{SME}} &= \Delta E_{c\leftrightarrow b}^{\bar{\text{H}},\,\text{SME}} \\
&= -(1 + \cos 2\theta_1)\, B_3^e - (1 - \cos 2\theta_1)\, B_3^p \\
&= -2\cos^2 \theta_1\, B_3^e - 2\sin^2 \theta_1\, B_3^p \, .
\end{aligned}
\tag{3.14}
$$

Comparing with the corresponding transition for hydrogen, and recalling that the hydrogen hyperfine states are spin-flipped relative to antihydrogen, we find

$$
\Delta E_{d\leftrightarrow a}^{\bar{\text{H}},\,\text{SME}} - \Delta E_{d\leftrightarrow a}^{\text{H},\,\text{SME}} = 4\cos^2 \theta_1\, b_3^e + 4\sin^2 \theta_1\, b_3^p \, ,
\tag{3.15}
$$

as first shown in [110].

This shows that at leading order in α, there is a predicted difference in the hydrogen and antihydrogen $1S$ hyperfine transitions proportional at zero magnetic field to the combination $b_3^e + b_3^p$ of SME couplings, while at high magnetic fields as in the ALPHA trap this dependence is on the electron coupling b_3^e alone. Recalling that the b_μ coupling in the SME Lagrangian is CPT odd, a difference between H and $\bar{\text{H}}$ in these transitions could therefore be interpreted as a signal of CPT violation. Assuming the energy levels described by (3.11) are the same for hydrogen and antihydrogen, and neglecting the $O(\alpha^2)$ contributions from other SME couplings, a measurement in ALPHA of the individual transition frequencies for $1S_d$–$1S_a$ or $1S_c$–$1S_b$ would therefore allow a bound to be placed on the CPT-violating electron coupling b_3^e. A corresponding measurement at zero magnetic field, as proposed by ASACUSA [121], would measure the combination $b_3^e + b_3^p$.

In fact, the published ALPHA measurement [36], with its quoted precision of 4×10^{-4}, is of the *difference* of the transition frequencies $\Delta E_{d\leftrightarrow a} - \Delta E_{c\leftrightarrow b}$, *not* the individual transition frequencies. The experimental procedure leading to this measurement is outlined above in Sect. 3.2.1. This has the advantage that the result is independent of the magnetic field B, as is evident from (3.11), so any uncertainty in determining the magnetic field that affects both resonances equally is not relevant. Unfortunately, from (3.14) we see that the dependence on the CPT violating parameters also cancels in this difference. This would at first sight imply that the difference of the hyperfine transition frequencies does not lead to a determination of the parameters $b_3^{e,p}$ and could not be considered as a test of CPT.

However, what we have shown is that even in the Lorentz and CPT violating SME theory, the difference of the hyperfine transition frequencies is still

$$\Delta E_{d \leftrightarrow a} - \Delta E_{c \leftrightarrow b} = \mathcal{E}_{HF}, \qquad (3.16)$$

i.e. the difference of the frequencies simply measures the hyperfine splitting $\mathcal{E}_{HF}$. Now, from (3.8), $\mathcal{E}_{HF}$ depends only on the mass and charge of the positron and antiproton, together with the g_e and g_p factors. As explained in Chap. 2 and Sect. 3.1, any difference in the mass or charge of a particle and its antiparticle leads to a violation of causality, and potentially unitarity or locality. These properties are far more fundamental than CPT symmetry, and their preservation is implicitly assumed in the formulation of the Lorentz and CPT violating SME effective theory.

This would, however, still leave the possibility that g_e or g_p could be different for the electron/positron or for the proton/antiproton as an explanation for a difference in the hyperfine splitting of antihydrogen and hydrogen. In this sense, the ALPHA measurement [36] can be viewed as a test of the identity of the anomalous magnetic moments for the particles and their antiparticles. Any difference would entail a violation of CPT, but is not predicted by the leading-order non-relativistic Hamiltonian derived from the SME.

$1S$–$2S$ Antihydrogen Transitions

Next we consider the $1S$–$2S$ transition in antihydrogen (blue arrows in Fig. 3.1). This is forbidden by the usual dipole selection rules for a single-photon transition and instead takes place through the much slower Doppler-free, two-photon transition. This gives rise to an extremely narrow spectral line and should permit precise tests of Lorentz and CPT symmetry in antihydrogen spectroscopy.

The derivation of the relevant $1S$ and $2S$ hyperfine energy levels follows that given above except that, as explained below, the leading correction to the $1S$–$2S$ transitions measured by ALPHA is $O(\alpha^2)$, so we need to consider both the E_{ij} and F_{kij} terms in the Hamiltonian (3.2), along with the higher-dimensional operator contribution $H_{SME}^{(5)}$ in (3.5).

First note that at zero magnetic field, where $\cos 2\theta_n = 0$, the hyperfine energy levels (3.13) can be written as

$$E_{SME} = A^e + A^p + \left(B_3^e + B_3^p \right) m_f, \qquad (3.17)$$

in the $|f\ m_f\rangle$ basis, independent of n. The two-photon selection rule $\Delta f = 0$, $\Delta m_f = 0$ therefore shows that these SME corrections give no contribution to the frequency $\Delta E_{2S \leftrightarrow 1S}$ of the $1S$–$2S$ transition. This is not, however, true in a magnetic field due to the n-dependence of the mixing angle θ_n in (3.10) and Footnote 2. Only in the high-field limit, where $\cos 2\theta_n \to 1$, does the $1S$–$2S$ transition frequency again become independent of the A and B_k terms in the SME Hamiltonian. Now consider the $O(p_i p_j)$ terms in the Hamiltonian (3.2). To compute their contribution to the hyperfine energy levels, we need to calculate expectation values of the form $\langle d|p_i p_j|d\rangle$, $\langle d|p_i p_j S_k^e|d\rangle$, etc. for all the states and for both the antiproton and positron spin operators.

At this point, we draw attention to the extremely detailed and clear calculation of the energy levels derived from H_{SME} for an arbitrary state $|n\ \ell\ j\ f\ m_f\rangle$ presented in

[114]. This paper derives the necessary Clebsch–Gordan coefficients for the evaluation of the expectation values of all the required combinations of the momentum and spin operators in these states. However, while this analysis is complete for calculating energy levels at zero magnetic field, we also require expectation values in the mixed hyperfine states, especially $|c\rangle$, to compute energy levels in strong magnetic fields such as those in the ALPHA trap. For these, we would also need *off-diagonal* matrix elements between states with differing f, m_f (see Footnote 2) in addition to those quoted in [114]. For states with arbitrary ℓ, j, f this is a substantial calculation.

Fortunately, the situation is enormously simplified for the $\ell = 0$ states, for which the wave function is isotropic, and it is then again straightforward to use the spin basis. Following [114], we first note

$$\langle n|p_i p_j|n\rangle = \frac{1}{3}\delta_{ij}\langle n|p^2|n\rangle = \frac{1}{3}\delta_{ij}\frac{\alpha^2 m_e^2}{n^2} \, , \tag{3.18}$$

which is independent of the spin of the state. (Recall the Bohr energy levels are $E_n = -\alpha^2 m_e/2n^2$.) The simplification is then clear. As an illustration, we can evaluate one of the expectation values contributing to E_{SME}^c in the same way as in (3.12), viz.

$$\begin{aligned}
\langle c|p_i p_j S_k^e|c\rangle &= -\frac{1}{2}\cos^2\theta_n \,\langle \Uparrow \downarrow \,|p_i p_j|\, \Uparrow \downarrow\rangle + \frac{1}{2}\sin^2\theta_n \,\langle \Downarrow \uparrow \,|p_i p_j|\, \Downarrow \uparrow\rangle \\
&= -\frac{1}{6}\delta_{ij}\left(\cos^2\theta_n \,\langle \Uparrow \downarrow \,|p^2|\, \Uparrow \downarrow\rangle - \sin^2\theta_n \,\langle \Downarrow \uparrow \,|p^2|\, \Downarrow \uparrow\rangle\right) \\
&= -\frac{1}{6}\delta_{ij}\cos 2\theta_n \frac{\alpha^2 m_e^2}{n^2} \, . \tag{3.19}
\end{aligned}$$

Ultimately, we find the energy levels for the trapped hyperfine states:

$$\begin{aligned}
E_{\mathrm{SME}}^d &= \tilde{A}^e + \tilde{A}^p - \left(B_3^e + B_3^p\right) \\
&\quad + \frac{1}{3}\frac{\alpha^2}{n^2}\,\mathrm{tr}_{i,j}\left(\tilde{E}_{ij}^e + \epsilon\tilde{E}_{ij}^p - \left(F_{ij3}^e + \epsilon F_{ij3}^p\right)\right) \tag{3.20}
\end{aligned}$$

and

$$\begin{aligned}
E_{\mathrm{SME}}^c &= \tilde{A}^e + \tilde{A}^p - \cos 2\theta_n \left(B_3^e - B_3^p\right) \\
&\quad + \frac{1}{3}\frac{\alpha^2}{n^2}\,\mathrm{tr}_{i,j}\left(\tilde{E}_{ij}^e + \epsilon\tilde{E}_{ij}^p - \cos 2\theta_n \left(F_{ij3}^e - \epsilon F_{ij3}^p\right)\right) \, , \tag{3.21}
\end{aligned}$$

where $\tilde{A} = A - a^{(5)}$ and $\tilde{E}_{ij} = E_{ij} - m_e^2 a_{ij}^{(5)}$ for both e and p. The c_{ij} traces simplify, since by suitable field redefinitions in the SME Lagrangian we are free to take $c^\mu{}_\mu = 0$.

The photon sector contributions are easily included using (3.6). Note that k_{AF} does not contribute here since we are considering $\ell = 0$ states. The k_F contribution is readily evaluated for these states using (3.18) and is proportional to the trace $\kappa_0 = (k_F)_{0i0i}$.

Substituting in terms of the original SME couplings, we therefore find:

$$
\begin{aligned}
E_{\text{SME}}^d = & -a_0^e - a^{(5)e} - m_e c_{00}^e - a_0^p - a^{(5)p} - m_e c_{00}^p \\
& + b_3^e + m_e d_{30}^e + H_{12}^e + b_3^p + m_e d_{30}^p + H_{12}^p \\
& - \frac{1}{3}\frac{\alpha^2}{n^2}\left[m_e^2\left(a_{ii}^{(5)e} + a_{ii}^{(5)p} \right) + \frac{5}{2}\left(m_e c_{00}^e + \epsilon m_p c_{00}^p \right) \right. \\
& \left. + 2m_e\kappa_0 + \left(b_3^e - \tilde{d}_3^e + \epsilon\left(b_3^p - \tilde{d}_3^p \right) \right) \right],
\end{aligned}
\tag{3.22}
$$

and

$$
\begin{aligned}
E_{\text{SME}}^c = & -a_0^e - a^{(5)e} - m_e c_{00}^e - a_0^p - a^{(5)p} - m_e c_{00}^p \\
& + \cos 2\theta_n \left(b_3^e + m_e d_{30}^e + H_{12}^e - b_3^p - m_e d_{30}^p - H_{12}^p \right) \\
& - \frac{1}{3}\frac{\alpha^2}{n^2}\left[m_e^2\left(a_{ii}^{(5)e} + a_{ii}^{(5)p} \right) + \frac{5}{2}\left(m_e c_{00}^e + \epsilon m_p c_{00}^p \right) \right. \\
& \left. + 2m_e\kappa_0 + \cos 2\theta_n \left(b_3^e - \tilde{d}_3^e - \epsilon\left(b_3^p - \tilde{d}_3^p \right) \right) \right],
\end{aligned}
\tag{3.23}
$$

where $\tilde{d}_3 = md_{03} + \frac{1}{2}md_{30} - \frac{1}{2}H_{12}$.[4]

This reproduces the relevant parts of Eq. (37) of [114] for the special case of $\ell = 0$ states in $\overline{\text{H}}$, and also includes the magnetic field dependence of the hyperfine state $|c\rangle$. Note that [114] includes also the e, f, g SME couplings which have been omitted here.

Finally, we can read off the SME contributions to the $1S_d$–$2S_d$ and $1S_c$–$2S_c$ transitions in antihydrogen, as measured by ALPHA [1] (see Fig. 3.1):

$$
\begin{aligned}
\Delta E_{2S_d \leftrightarrow 1S_d}^{\overline{\text{H}}} = & \frac{1}{4}\alpha^2\left[m_e^2\left(a_{ii}^{(5)e} + a_{ii}^{(5)p} \right) + 2m_e\kappa_0 \right. \\
& \left. + \frac{5}{2}\left(m_e c_{00}^e + \epsilon m_p c_{00}^p \right) + \left(b_3^e - \tilde{d}_3^e + \epsilon b_3^p - \epsilon \tilde{d}_3^{\ p} \right) \right],
\end{aligned}
\tag{3.24}
$$

---

[4]To compare with [103, 115] (see also [122]), these references rewrite the SME tensor couplings in a spherical basis then, after making appropriate field redefinitions, consider the isotropic components which are sufficient to describe the $\ell = 0$ states. In their notation, with couplings a_{njm}^{NR}, the dictionary to compare with (3.22) and (3.23) is then $a_{000}^{\text{NR}} = a_0 + a^{(5)}$ and $a_{200}^{\text{NR}} = \frac{1}{3}\text{tr}\, a_{ij}^{(5)}$. The c_{00} contributions are similarly written as c_{200}^{NR}. Later, when we consider the $1S$–$2P$ transitions, we find a further dependence on the combination $a_Q^{(5)} = a_{11}^{(5)} + a_{22}^{(5)} - 2a_{33}^{(5)}$ of the $a_{ij}^{(5)}$, which is proportional to the non-isotropic couplings a_{220}^{NR}. Similarly for c_Q.

and

$$\Delta E^{\bar{H}}_{2S_c \leftrightarrow 1S_c} = (\cos 2\theta_2 - \cos 2\theta_1) \left(b_3^e + m_e d_{30}^e + H_{12}^e - b_3^p - m_e d_{30}^p - H_{12}^p \right)$$
$$+ \frac{1}{4} \alpha^2 \left[m_e^2 \left(a_{ii}^{(5)e} + a_{ii}^{(5)p} \right) + 2 m_e \kappa_0 + \frac{5}{2} \left(m_e c_{00}^e + \epsilon m_p c_{00}^p \right) \right.$$
$$\left. - \frac{1}{3} (\cos 2\theta_2 - 4 \cos 2\theta_1) \left(b_3^e - \tilde{d}_3^e - \epsilon b_3^p + \epsilon \tilde{d}_3^{\,p} \right) \right].$$
$$(3.25)$$

Although $\cos 2\theta_n \simeq 1$ at the high magnetic field in the ALPHA trap, we have kept this dependence here. It shows an important feature, viz. that unlike the $|d\rangle \to |d\rangle$ transition, the $|c\rangle \to |c\rangle$ transition has a contribution at $O(1)$ in the fine structure constant, albeit highly suppressed by a magnetic field factor. We come back to this below. At zero field, the $\tilde{d}_3$ and b_3 terms do not contribute to the E^c_{SME} energy levels, both then being proportional to m_f. In either case, we see from (3.24) and (3.25) that the SME contributions to the c and d transitions are different.

To compare with the $1S$–$2S$ transitions in hydrogen, again recalling that the corresponding states are spin-flipped, we make the CPT conjugation sign changes on the relevant SME couplings as described above and find:

$$\Delta E^{\bar{H}}_{2S_d \leftrightarrow 1S_d} - \Delta E^{H}_{2S_d \leftrightarrow 1S_d} = \frac{1}{2} \alpha^2 \left[m_e^2 \left(a_{ii}^{(5)e} + a_{ii}^{(5)p} \right) + b_3^e + \epsilon b_3^p \right], \qquad (3.26)$$

and

$$\Delta E^{\bar{H}}_{2S_c \leftrightarrow 1S_c} - \Delta E^{H}_{2S_c \leftrightarrow 1S_c} = 2 (\cos 2\theta_2 - \cos 2\theta_1) \left(b_3^e - b_3^p \right)$$
$$+ \frac{1}{2} \alpha^2 \left[m_e^2 \left(a_{ii}^{(5)e} + a_{ii}^{(5)p} \right) - \frac{1}{3} (\cos 2\theta_2 - 4 \cos 2\theta_1) \left(b_3^e - \epsilon b_3^p \right) \right].$$
$$(3.27)$$

Clearly, this only depends on the CPT odd couplings in the SME Lagrangian. Again note the $O(1)$ contribution in the $|c\rangle \to |c\rangle$ transition only. We comment on the significance of these results on $1S$–$2S$ spectroscopy for fundamental physics below, after first considering other transitions accessible to the ALPHA programme.

$1S$–$2P$ and Other Antihydrogen Transitions

ALPHA have also recently measured the $1S$–$2P$ transition in antihydrogen [28, 106] (green arrows in Fig. 3.1), the first involving a state with non-zero orbital angular momentum. As such, it has some extra interest from a fundamental physics perspective since it is directly sensitive to the potential spin-independent CPT violation (3.6) in the photon sector, parametrised in the SME by the effective coupling k_{AF}.

In the absence of an external magnetic field, the $2P$ states are split by the spin-orbit coupling into a $j = 3/2$ quartet and a $j = 1/2$ doublet, with energy difference $\mathcal{E}_{FS}$. With non-zero B, Zeeman splitting removes the remaining degeneracy with respect to m_j, with the $m_j = 1/2$ states with $j = 3/2, 1/2$ being mixed and similarly for the $m_j = -1/2$ states. The first step is therefore to determine these energy eigenstates and mixing angles for the magnetic fields present in the ALPHA trap.

In this case, it is a good approximation to neglect the hyperfine splitting, which is relatively small for the $2P$ states, and include the $m_I = \pm 1/2$ antiproton spin only after finding the positron eigenstates. The effective Hamiltonian, including the spin-orbit coupling, is then simply

$$H_{SO} = \frac{2}{3}\mathcal{E}_{FS}\, \mathbf{L}.\mathbf{S} - \mu_B\, (\mathbf{L} + g_e\mathbf{S})\,.\,\mathbf{B}$$
$$= \frac{1}{3}\mathcal{E}_{FS}\left(J^2 - L^2 - S^2\right) - \mu_B\,(L_z + g_e S_z)\,B \tag{3.28}$$

Neither $|n\,\ell\,s\,j\,m_j\rangle$ nor $|n\,\ell\,s\,m_\ell\,m_s\rangle$ states are eigenstates of H_{SO} and either basis can be used to describe the mixed states at non-zero B. Since the magnetic field in ALPHA is relatively high, and with an eye to the inclusion of SME couplings, we find it more convenient to describe the states in the $|m_\ell\,m_s\rangle$ basis. Note also that $m_j = m_\ell + m_s$ is a good quantum number for H_{SO}, since $[H_{SO}, J_z] = 0$, but j is not. This selects the mixed states as described above and, after diagonalising the Hamiltonian, we find the following $2P$ eigenstates and corresponding energy eigenvalues in the $|m_\ell\,m_s\rangle$ basis:

$$
\begin{aligned}
|a\rangle &= |-1\ -\tfrac{1}{2}\rangle \\
|b\rangle &= \cos\psi\,|-1\ \tfrac{1}{2}\rangle + \sin\psi\,|0\ -\tfrac{1}{2}\rangle \\
|c\rangle &= \sin\eta\,|1\ -\tfrac{1}{2}\rangle + \cos\eta\,|0\ \tfrac{1}{2}\rangle \\
|d\rangle &= |1\ \tfrac{1}{2}\rangle \\
|e\rangle &= -\sin\psi\,|-1\ \tfrac{1}{2}\rangle + \cos\psi\,|0\ -\tfrac{1}{2}\rangle \\
|f\rangle &= \cos\eta\,|1\ -\tfrac{1}{2}\rangle - \sin\eta\,|0\ \tfrac{1}{2}\rangle
\end{aligned}
\tag{3.29}
$$

where

$$\tan\psi = \frac{1}{2\sqrt{2}\mathcal{E}_{FS}}\left(\mathcal{E}_{FS} + 3\mu_B B + 6\mathcal{E}_1(B)\right), \tag{3.30}$$

and

$$\tan\eta = \frac{1}{2\sqrt{2}\mathcal{E}_{FS}}\left(-\mathcal{E}_{FS} + 3\mu_B B + 6\mathcal{E}_1(-B)\right), \tag{3.31}$$

with

$$\mathcal{E}_1(B) = \tfrac{1}{2}\left[\mathcal{E}_{FS}^2 + \tfrac{2}{3}\mu_B B\,\mathcal{E}_{FS} + (\mu_B B)^2\right]^{1/2} \tag{3.32}$$

and we have set $g_e = 2$.[5]

The corresponding energy levels of H_{SO} are:

$$E_a = \frac{1}{3}\mathcal{E}_{FS} + 2\mu_B B$$

$$E_b = -\frac{1}{6}\mathcal{E}_{FS} + \frac{1}{2}\mu_B B + \mathcal{E}_1(B)$$

$$E_c = -\frac{1}{6}\mathcal{E}_{FS} - \frac{1}{2}\mu_B B + \mathcal{E}_1(-B)$$

$$E_d = \frac{1}{3}\mathcal{E}_{FS} - 2\mu_b B$$

$$E_e = -\frac{1}{6}\mathcal{E}_{FS} + \frac{1}{2}\mu_B B - \mathcal{E}_1(B)$$

$$E_f = -\frac{1}{6}\mathcal{E}_{FS} - \frac{1}{2}\mu_B B - \mathcal{E}_1(B). \tag{3.33}$$

These are illustrated in green in the upper panel of Fig. 3.1, with reference to the $2S$ level including the Lamb shift. Clearly, for zero magnetic field, $E_{a,...d} = \frac{1}{3}\mathcal{E}_{FS}$, and $E_{e,f} = -\frac{2}{3}\mathcal{E}_{FS}$, consistent with their interpretation as the $j = 3/2$ and $j = 1/2$ states respectively. The corresponding limits for the mixing angles are $\tan\psi = \sqrt{2}$ and $\tan\eta = 1/\sqrt{2}$, which reproduce the required Clebsch–Gordan factors in (3.29) to convert between $|j\ m_j\rangle$ and $|m_\ell\ m_s\rangle$ states.

For very large fields, the mixing angles both go to the limit $\pi/2$, and the limiting form of the states can be immediately read off from (3.29). For these $2P$ states, however, the ALPHA magnetic field $B = 1.03$ T does not fully reach this limit. In fact, at this value of B, the mixing angles are $\tan\psi \simeq 3.76$ and $\tan\eta \simeq 2.49$. These imply the following values which we need below to parametrise the contributions from Lorentz and CPT violating parameters, $\cos 2\psi = -0.868$ and $\cos 2\eta = -0.721$.

Finally, we include the antiproton spin by simply taking the direct product with each of these states, with notation $|a+\rangle = |a\rangle\,|\Uparrow\rangle$, $|a-\rangle = |a\rangle\,|\Downarrow\rangle$, etc.

The transitions of interest to us here are $1S_d$–$2P_{c-}$, $1S_d$–$2P_{f-}$, $1S_c$–$2P_{c+}$ and $1S_c$–$2P_{f+}$, with the notation in (3.10) for the $1S$ hyperfine states. Of these, the first two have recently been measured by ALPHA [106], with the hyperfine states resolved. To find the contribution to these transition frequencies from the Lorentz and CPT couplings, we first need to find the expectation value of the effective SME Hamiltonian (3.2) in these $2P$ states. Evidently, there is a contribution already at $O(1)$ in the fine structure constant, but we shall give a complete result up to $O(\alpha^2)$ including also the photon sector couplings.

---

[5]These results agree with those in [120], up to a different choice of phase (sign) for the $|e\rangle$ state and an alternative definition of the mixing angle for $|b\rangle$, $|e\rangle$. With our choice, $|e\rangle$ becomes the $|j = 1/2,\ m_j = 1/2\rangle$ state at $B = 0$.

Technically, the new feature arising with the $2P$ states is that since they have a non-zero angular momentum, the wave functions are no longer isotropic and we cannot use the simplification (3.18) for the expectation values $\langle \hat{p}_i \hat{p}_j \rangle$. To overcome this [114], we first express $\hat{p}_i \hat{p}_j$ in a spherical tensor basis, defining coefficients C_{ij}^M from the expansion in spherical harmonics,

$$\hat{p}_i \hat{p}_j - \frac{1}{3}\delta_{ij} = \sum_{M=-2}^{2} C_{ij}^M Y_2^M(\theta, \phi) . \tag{3.34}$$

Matrix elements in an $|n \; \ell \; m_\ell\rangle$ basis are then found using a well-known formula for the product of three spherical harmonics in terms of Clebsch–Gordan coefficients. The result is,

$$\langle n \; \ell \; m_\ell | \hat{p}_i \hat{p}_j | n \; \ell \; m'_\ell \rangle = \frac{1}{3} \delta_{ij} \, \delta_{m_\ell m'_\ell}$$
$$+ \sqrt{\frac{5}{4\pi}} \, C_{ij}^{m_\ell - m'_\ell} \langle \ell \; 0 ; 2 \; 0 | \ell \; 0 \rangle \langle \ell \; m'_\ell ; 2 \; m_\ell - m'_\ell | \ell \; m_\ell \rangle . \tag{3.35}$$

One of the simplifications of expressing the $2P$ Zeeman states in the $|m_\ell \; m_s\rangle$ basis is that we only need the diagonal matrix elements in (3.35). In this case, specialising to $\ell = 1$ and $m_\ell = m'_\ell$, we have

$$\langle 2 \; 1 \; m_\ell | \hat{p}_i \hat{p}_j | 2 \; 1 \; m_\ell \rangle = \frac{1}{3} \delta_{ij} - \sqrt{\frac{1}{2\pi}} \, C_{ij}^0 \langle 1 \; m_\ell ; 2 \; 0 | 1 \; m_\ell \rangle . \tag{3.36}$$

The relevant coefficients are $C_{ij}^0 = -\sqrt{\frac{4\pi}{45}} \delta_{ij}^Q$ with $\delta_{ij}^Q = \delta_{i1}\delta_{j1} + \delta_{i2}\delta_{j2} - 2\delta_{i3}\delta_{j3}$, so finally we find the required expectation value,

$$\langle 2 \; 1 \; m_\ell | \hat{p}_i \hat{p}_j | 2 \; 1 \; m_\ell \rangle = \frac{1}{3} \delta_{ij} + \frac{1}{3}\sqrt{\frac{2}{5}} \langle 1 \; m_\ell ; 2 \; 0 | 1 \; m_\ell \rangle \delta_{ij}^Q . \tag{3.37}$$

Since the individual $2P$ states in (3.29) are themselves eigenstates of L_z and S_z^e, it is now straightforward to evaluate all the required matrix elements such as $\langle \hat{p}_i \hat{p}_j \rangle$, $\langle \hat{p}_i \hat{p}_j S_z^e \rangle$, etc. in these states. The photon sector contributions from (3.6) are also readily incorporated using these matrix elements. Here, we simply quote the final results for the transition frequencies for $1S_d$–$2P_{c_-}$ and $1S_d$–$2P_{f_-}$. For $1S_d$–$2P_{c_-}$, we have[6]

---

[6]Note that for simplicity we have set $\cos 2\theta_1 \to 1$ here, since for the ALPHA magnetic field, $\cos 2\theta_1 \simeq 0.998$.

$$\Delta E^{\bar{H}}_{2P_{c-}\leftrightarrow 1S_d} = -(1+\cos 2\eta)\left(b^e_3 + m_e d^e_{30} + H^e_{12}\right)$$

$$+ \frac{1}{4}\alpha^2\left[-m_e^2\left(a^{(5)e}_{ii} + a^{(5)p}_{ii}\right) - \frac{1}{30}m_e^2(1+3\cos 2\eta)\left(a^{(5)e}_Q + a^{(5)p}_Q\right)\right.$$

$$+ \frac{5}{2}\left(m_e c^e_{00} + \epsilon m_p c^p_{00}\right) + \frac{1}{30}(1+3\cos 2\eta)\left(m_e c^e_Q + \epsilon m_p c^p_Q\right)$$

$$+ \frac{1}{3}(4+\cos 2\eta)\left(b^e_3 - \tilde{d}^e_3\right) + \frac{1}{3}\epsilon\left(b^p_3 - \tilde{d}^p_3\right)$$

$$- \frac{1}{30}(3+\cos 2\eta)\left(b^e_3 + 2\tilde{d}^e_3\right) + \frac{1}{30}\epsilon(1+3\cos 2\eta)\left(b^p_3 + 2\tilde{d}^p_3\right)$$

$$\left. + 2m_e\kappa_0 - \frac{1}{30}(1+3\cos 2\eta)\,m_e\kappa_Q + \frac{1}{2}(1-\cos 2\eta)\,(k_{AF})_3 \right],$$

$$\tag{3.38}$$

where $a^{(5)}_Q = a^{(5)}_{ij}\delta^Q_{ij}$, $c_Q = c_{ij}\delta^Q_{ij}$ and $\kappa_Q = (k_F)_{0i0j}\delta^Q_{ij}$. Recall (Footnote 4) that these non-isotropic couplings $a^{(5)}_Q$, c_Q are proportional to the couplings denoted a^{NR}_{220}, c^{NR}_{220} respectively in [103, 115]. An identical result holds for $\Delta E^{\bar{H},SME}_{2P_{f-}\leftrightarrow 1S_d}$ with the substitution $\cos 2\eta \to -\cos 2\eta$ throughout, and similar expressions can be derived for transitions involving $1S_c$. Experimentally, while inhomogenous broadening of the $1S$–$2P$ line initially obscured the hyperfine structure [28], state selectivity has since been achieved by ejecting the unwanted $1S_c$ hyperfine population from the trap before spectroscopy begins [106].

Notice that for the magnetic field in ALPHA, both transitions have $O(1)$ contributions from the SME couplings dependent on the positron spin, viz. b^e_3, d^e_{30} and H^e_{12}, but not (in this approximation) from the corresponding antiproton couplings. At still larger magnetic fields, where $\cos 2\eta \to -1$, these contributions would cancel out in the $1S_d$–$2P_{c-}$ transition only. Also recall that at zero field, $\cos 2\eta = 1/3$.

The other qualitatively new feature of (3.38) compared to the $1S$–$2S$ transition is in the photon sector, where the $1S$–$2P$ transitions become sensitive to the CPT violating coupling $(k_{AF})_3$, as well as an independent combination κ_Q of the CPT even $(k_F)_{0i0j}$ couplings. Note also the appearance of the new Lorentz violating combination c^e_Q in the positron sector.

As always, to expose the difference with the corresponding transitions in hydrogen, we keep only the CPT odd couplings, viz. $a^{(5)e,p}_{ij}$, $b^{e,p}_3$ and $(k_{AF})_3$. This leaves the much simpler formula (approximating $\epsilon = 0$ here),

$$\Delta E^{\bar{H}}_{2P_{c-}\leftrightarrow 1S_d} - \Delta E^{H}_{2P_{c-}\leftrightarrow 1S_d} = -2(1+\cos 2\eta)\,b^e_3$$

$$+ \frac{1}{2}\alpha^2\left[-m_e^2\left(a^{(5)e}_{ii} + a^{(5)p}_{ii}\right) - \frac{1}{30}m_e^2(1+3\cos 2\eta)\left(a^{(5)e}_Q + a^{(5)p}_Q\right)\right.$$

$$\left. + \frac{1}{3}(4+\cos 2\eta)b^e_3 - \frac{1}{30}(3+\cos 2\eta)b^e_3 + \frac{1}{2}(1-\cos 2\eta)(k_{AF})_3 \right],$$

$$\tag{3.39}$$

with the corresponding result for $1S_d$–$2P_{f-}$. Note that, unlike the $1S$–$2S$ transitions, the $1S$–$2P$ transition frequencies also involve the non-isotropic $a_Q^{(5)}$ SME couplings.

The ALPHA programme involves a detailed study of a variety of further transitions, outlined in [38]. Similar theoretical methods can be applied to determine the dependence on the Lorentz and CPT violating couplings for all the relevant eigenstates, using the key formula (3.35) with the appropriate Clebsch–Gordan coefficients and taking account of the magnetic field dependence of the mixing angles amongst the Zeeman states.

Testing Lorentz and CPT Symmetry

We close this section with a general discussion of the implications of the ALPHA measurements of the $1S$ hyperfine, $1S$–$2S$ and $1S$–$2P$ antihydrogen transitions for the effective theory of Lorentz and CPT violation. Conversely, we comment on how the general features of this theory may motivate future measurements in antihydrogen spectroscopy.

The first point to emphasise is that the precision reached experimentally in, for example, the $1S$–$2S$ transition frequency far exceeds that possible from a first principles QED calculation. This means that to establish any violations of fundamental principles such as Lorentz and CPT symmetry we need to *compare* measurements. In the case of CPT, this means comparing the spectrum of H and $\overline{\text{H}}$ in sufficiently similar environments that we can control systematics to high precision. Such *instantaneous* comparisons of the H and $\overline{\text{H}}$ spectra will be different if and only if CPT is broken. For Lorentz invariance, H (or $\overline{\text{H}}$) transition frequencies should be compared at different times, to look for possible sidereal or annual variations. A detailed analysis of the formalism required to compare measurements made in the laboratory frame with a sidereal frame, or the boosted frame reflecting the Earth's annual motion around the sun, is beyond the scope of this article. A complete treatment relevant to H and $\overline{\text{H}}$ spectroscopy is given in [115].

Second, in comparing sensitivities to the SME couplings from different transitions, we need to distinguish between the *relative* precision of the measurement and the *absolute* precision of the energy sensitivity, which bounds both the dimensional SME couplings such as b_3 and the dimensionless couplings such as c_{00} (which arise in energy levels accompanied by factors of m_e).

A further important point, which is evident from our explicit expressions for the atomic energy levels and transitions, is that the Lorentz violating couplings always appear in combinations comprising CPT even and CPT odd couplings. For example, the $1S$ hyperfine transitions (3.14) involve the combination $B_3^e = -b_3^e - m_e d_{30}^e - \frac{1}{2}\epsilon_{kij}H_{ij}^e$, arising directly from the Hamiltonian (3.2), where b_3 is CPT odd while d_{30} and H_{12} are CPT even. Similarly, the $1S$–$2S$ transition involves the CPT even c_{00} together with the higher-dimensional CPT odd coupling $\text{tr } a_{ij}^5 \sim a_{200}^{\text{NR}}$. As noted in [115], this is a very general feature of the SME effective theory. In practice, this means that a search for Lorentz violation from an experiment with matter alone can only bound this combination. It cannot *on its own* test for CPT violation, since there remains the possibility of a cancellation of the CPT violating couplings (e.g. b_3) and the CPT even couplings (e.g. d_{30} and H_{12}). Again, this shows that to identify unambiguously a signal for CPT violation, we need to compare experiments on equivalent matter and antimatter systems.

In essence, this is the same idea we have already tried to exploit in Sect. 2.6, where we considered potential new background fields and the equivalence principle. Here, we similarly entertain the possibility of a *cancellation* between the Lorentz violating couplings within a pure matter system, while allowing them to add to give an observable effect for the equivalent pure antimatter system.

Now, focusing first on CPT, comparing the $1S$ hyperfine spectrum (3.15) of H and $\overline{\text{H}}$ at zero magnetic field allows a bound to be placed on the combination $b_3^e + b_3^p$, while at the ALPHA magnetic field the sensitivity is essentially to b_3^e alone. Combining these, we could bound both b_3^e and b_3^p (always recalling that b_3^p and related quantities is an effective parameter for the QCD bound state proton, not actually a parameter in the SME Lagrangian itself). This would, however, require individual measurements of the $1S_d$–$1S_a$ and $1S_c$–$1S_b$ transitions, whereas the published ALPHA results [36] are for the difference alone, with a quoted precision of 4×10^{-4}. If, for illustration, we assume a similar precision could be reached for the individual transition frequencies, then since the frequency of the $1S_d$–$1S_a$ hyperfine transition is 29 GHz, this would correspond to an absolute energy precision of 12 MHz or 4.8×10^{-8} eV and would imply a bound on $|b_3^e| \lesssim 10^{-17}$ GeV.

A similar comparison for the $1S_d$–$2S_d$ transition is given in (3.26). Here, the leading contribution of the CPT violating SME couplings arises only at $O(\alpha^2)$, and depends on $b_3^{e,p}$ and the higher-dimensional $a^{(5)}$ couplings, whose contribution is expected to be relatively suppressed by $O(m_e^2/\Lambda^2)$. The b_3^p contribution is accompanied by the mass suppression factor $\epsilon = m_e^2/m_p^2 \sim 10^{-6}$. So if first we interpret (3.26) as bounding b_3^e, the ALPHA precision of 2×10^{-12} on the $1S_d$–$2S_d$ transition frequency of 2.466×10^{15} Hz [2] gives a bound $|b_3^e| \lesssim 7 \times 10^{-16}$ GeV. This illustrates the point raised above, that a higher relative precision measurement of a higher frequency spectral line can nevertheless result in a less stringent bound on the CPT violating coupling b_3^e.

On the other hand, if we impose the existing bounds on b_3^e quoted in [44],[7] then the $1S_d$–$2S_d$ measurement can be used to give a bound on the sum of the higher-dimensional couplings $a^{(5)}$ for e and p, viz. $\text{tr}\, a_{ij}^{(5)} \sim a_{200}^{\text{NR}} \lesssim 10^{-9}$ GeV^{-1}.

The analysis differs for the $1S_c$–$2S_c$ transitions in H and $\overline{\text{H}}$. In this case, from (3.27) we have a contribution proportional to b_3^e at $O(1)$ but suppressed by the magnetic field dependent factor $(\cos 2\theta_2 - \cos 2\theta_1) = 1.2 \times 10^{-3}$ at the ALPHA trap magnetic field of 1.03 T. This is greater than the $O(\alpha^2)$ contribution discussed above. If the same precision can be attained as for the $1S_d$–$2S_d$ transition, this would give a bound $|b_3^e| \lesssim 10^{-17}$ GeV. This is comparable with the illustrative bound given above for a potential determination of b_3^e from the hyperfine transitions.

---

[7]Note that the formula quoted in [115, 122] for the SME contribution to the $1S$–$2S$ transition omits the spin-dependent couplings such as b_3 in (3.26). This is because they assume prior to calculating the energy levels that these couplings are negligible compared to a_{200}^{NR} and c_{00}, on the basis of the existing experimental constraints given in [44].

The b_3^e coupling can also be bounded from the $1S$–$2P$ transitions. Temporarily neglecting the photon coupling $(k_{AF})_3$ in (3.39), along with the $a_{ij}^{(5)}$ couplings, and using the ALPHA precision of 76 MHz for the resolved $1S_d$–$2P_{c_-}$ transition [106] gives a bound $|b_3^e| \lesssim 5 \times 10^{-16}$ GeV.

Conversely, imposing the bound [44] on b_3^e, and assuming the $1S$–$2S$ bound on $a_{ii}^{(5)}$, would enable a bound $a_Q^{(5)} \lesssim 10^{-3}$ GeV^{-1} to be placed on the non-isotropic higher-dimensional couplings.

As emphasised above, however, a new feature of $1S$–$2P$, and any transition involving states with non-zero orbital angular momentum, is its sensitivity to a potential spin-independent CPT violation arising in the photon sector. Accepting the quoted bounds on b_3^e in [44], the result (3.39) bounds the CPT violating photon coupling $|(k_{AF})_3| \lesssim 10^{-11}$ GeV. This is many orders of magnitude below the bound $|k_{AF}| \lesssim 10^{-42}$ GeV [44] deduced from astronomical observations from gamma ray bursts or the CMB. Nevertheless, the comparison of the ALPHA $1S$–$2P$ result with hydrogen gives an interesting illustration of limiting CPT violation in the photon sector in an atomic physics experiment.

Next, we consider how these spectral transitions in H and $\overline{\text{H}}$ may test for violations of Lorentz symmetry. Of course, all the SME couplings violate Lorentz invariance but we focus first on the CPT even ones. Of particular interest are the spin-independent c_{00} couplings.[8] These do not contribute to the $1S$ hyperfine transitions, so the best bounds arise from sidereal or annual variations in the measured $1S$–$2S$ frequency. Currently, the most stringent bound comes from high precision hydrogen spectroscopy [109], which now allows the $1S$–$2S$ frequency to be measured to 10^{-15}.

Extracting a bound on the c_{00} coefficients from the null observation of annual variations in the hydrogen $1S$–$2S$ transition is not straightforward and we refer to [109, 113, 115] for details. The essential point is that comparing frames of reference at different points in the Earth's orbit introduces a suppression in the implied bound by a boost factor of $O(v_E/c) \sim 10^{-4}$, where v_E is the Earth's orbital velocity. This is the origin of the bound of order $|c_{00}| \lesssim 10^{-11}$ quoted in [109]. However, on its own this measurement only limits the combination of c_{00} with the CPT odd operator a_{200}^{NR}. To isolate c_{00}, we need an independent bound on a_{200}^{NR}, which requires a measurement of the corresponding transition in antihydrogen. In fact, the above bound $a_{200}^{\text{NR}} \lesssim 10^{-9}$ GeV^{-1} deduced from ALPHA is just sufficient to justify, a posteriori, this bound on c_{00}.

Finally, to compare the potential sensitivity of the antihydrogen bounds on b_3^e with existing results, we note again that measurements on purely matter systems, e.g. on sidereal variations in the spin precession frequency of electrons in a Penning trap (see [123] for a summary), can only bound the combination of B_3^c (the charge conjugate of B_3 as defined here). These bounds (see under $|\tilde{b}_{X,Y,Z}|$ in [44]) are typically of $O(10^{-23})$ though can be significantly lower from torsion pendulum

---

[8]In principle, the independent role of all the Lorentz violating couplings, including the spin-dependent $\tilde{d}_{30}$ and H_{12}, may be extracted from the frequencies of the variety of transitions considered here, especially given their different magnetic field dependence.

experiments. However, the only bound quoted in [44] on $|\underline{b}^e|$ itself comes from the comparison of spin precession frequencies of electrons and positrons [124], with the result $|\underline{b}^e| \lesssim 10\,\mathrm{Hz}$, equivalent to $4 \times 10^{-22}\,\mathrm{GeV}$.

It is interesting to compare the interpretation of these measurements of the anomalous magnetic moment for the electron/positron [124] and proton/antiproton [125–128] in a Penning trap with the corresponding determination from the hyperfine splitting in H/$\overline{\mathrm{H}}$. Penning trap measurements compare the spin precession (Larmor) and cyclotron frequencies ω_s and ω_c in a background magnetic field, the difference being the 'anomalous' frequency $\omega_a = \omega_s - \omega_c$. In a conventional theory, this measures the $g - 2$ factor for the test particle.[9] In the SME, however, the CPT odd b_3 coefficient in the Hamiltonian (3.2) modifies the spin precession frequency [129] and to leading order,

$$\frac{\omega_a}{\omega_c} = \frac{1}{2}(g - 2) - \frac{2m}{eB} b_3 . \tag{3.40}$$

A possible CPT-violating difference in the ratio ω_a/ω_c for, say, the electron and positron could then be attributable either to a difference in the g factors for the particle and antiparticle, which depends on quantum loop corrections in QED, or to the direct Lorentz and CPT violating b_3 coefficient in the SME. The bounds on $b_3^{e,p}$ quoted in [124] and [125–128] are subject to the assumption that g^e and g^p are the same for the particle and antiparticle.

In contrast, while the individual (anti)hydrogen hyperfine transitions depend at leading order on the SME coefficients $b_3^{e,p}$, as we showed in (3.16) this dependence cancels out in the difference of the $1S_d$–$1S_a$ and $1S_c$–$1S_b$ transitions. The ALPHA [36] measurement of the hyperfine splitting $\mathcal{E}_{HF}$ depends purely on the anomalous magnetic moments of the positron and antiproton given by the corresponding g factors alone. Comparison of (3.40) with (3.8) for $\mathcal{E}_{HF}$ then makes it clear that the e^-/e^+ and $p/\overline{p}$ Penning trap and H/$\overline{\mathrm{H}}$ hyperfine splitting measurements provide *complementary* tests of CPT invariance.

To summarise, setting aside the details of the SME parametrisation, the results above make it clear that Lorentz and CPT violation can arise in subtly different ways in all the antihydrogen spectral transitions and, in the event of a non-null discovery, many measurements may be necessary to pin down the origin of CPT breaking. Sidereal and annual variations can also place competitive bounds on Lorentz violation. Moreover, in terms of looking for radically new physics, we should not lose sight of the fact that the SME is itself in some sense conservative, being a conventional effective quantum field theory built in the standard way from causal fields in rep-

---

[9]The cyclotron frequency is $\omega_c = eB/m$ while the spin precession frequency, which depends on the magnetic moment, is $\omega_s = g\mu_B B$. The difference therefore gives

$$\frac{\omega_a}{\omega_c} \equiv \frac{\omega_s - \omega_c}{\omega_c} = \frac{1}{2}(g - 2) .$$

However, in the SME, the spin-dependent b_3 term in the Hamiltonian (3.2) gives an extra contribution to the spin precession frequency alone, so that $\omega_s = g\mu_B B - 2b_3$. In this theory, the ratio ω_a/ω_c therefore has the additional, B-dependent, factor shown in (3.40).

resentations of the Lorentz group. All this further motivates the most extensive and high precision analysis of the whole antihydrogen spectrum, including the search for sidereal and annual variations.

3.2.3 New Background Fields

Until now, we have considered the possibility of new, long-range background fields ('fifth forces') from the perspective of their gravity-like effects on weak equivalence principle experiments (see Sect. 2.6). Here, we point out an interesting effect of a long-range $U(1)_{B-L}$ interaction on atomic spectroscopy and show how this can limit the allowed coupling strength to high precision.

The idea is that in a $U(1)_{B-L}$ gauge theory in which the gauge boson is essentially massless (strictly, with a sufficiently small mass $m_{Z'} < 10^{-14}$ eV that the force has a range greater the Earth's radius), the Earth itself acts as a source creating the $U(1)_{B-L}$ analogue of an electric field at the surface of magnitude $\mathcal{E}_{B-L} = Q_{B-L}^{Earth} g'/4\pi R_E^2$. Here, Q_{B-L}^{Earth} is the $B - L$ charge of the Earth (the number of neutrons) and R_E is its radius. A hydrogen atom placed in this field acts as a $U(1)_{B-L}$ dipole since the proton has $Q_{B-L}^p = 1$ while the electron has $Q_{B-L}^e = -1$.

The atom therefore experiences a $U(1)_{B-L}$ analogue of the Stark effect, which occurs when an atom is placed in a conventional electric field. This produces a $U(1)_{B-L}$ linear Stark shift in the $n = 2$ energy levels given by adapting the usual formula, viz.

$$\Delta E \simeq \pm g' \mathcal{E}_{B-L} a_0 , \tag{3.41}$$

where a_0 is the Bohr radius. That is,

$$\Delta E \simeq Q_{B-L}^{Earth} \alpha' a_0/R_E^2 \simeq 10^{21} \alpha' \text{ eV} . \tag{3.42}$$

This Stark shift would be opposite in sign for antihydrogen because of the opposite $B - L$ charge of the antiparticles, so in the absence of any other non-standard model interactions, a comparison of the H and $\overline{\text{H}}$ spectra would reveal the shift (3.42). From the current absolute energy sensitivity of the antihydrogen 1S–2S transition of 2×10^{-20} GeV [2] we can therefore place a bound $\alpha' \lesssim 10^{-33}$ on the $U(1)_{B-L}$ coupling.

This is a remarkable level of precision at which to bound a new gauge coupling. Nevertheless, as noted in Sect. 2.6, existing equivalence principle experiments already constrain the coupling to the far lower value $\alpha' < 10^{-49}$, the extremely small value of course reflecting the huge $B - L$ charge of the Earth. Despite the high precision of atomic spectroscopy, it therefore seems that gravitational tests remain a better test of new long-range 'fifth force' interactions.

3.3 Antihydrogen and Gravity

There are currently three initiatives aimed at direct investigations of antimatter gravity based upon free fall measurements [130–132]. We briefly describe these below, noting that the ALPHA-g experiment [132] has developed using experience and knowledge gained from the trajectory analyses described in Sect. 3.1. This was also to the fore in the gravity investigation described by Amole and coworkers [34] in which, in essence, the equation of motion (3.1) was used, but assuming the antihydrogen "charge anomaly" $Q = 0$. The ALPHA apparatus [24] is a horizontal antihydrogen trapping device (which as described in [32] is not optimum for gravity investigations), and the annihilations of the anti-atoms escaping from the trap as the magnetic holding fields were lowered were used to deduce rough limits, mostly as a demonstration of proof-of-principle, on the aforementioned parameter F. Values of $-65 < F > 110$ were excluded at a level of 95% statistical significance.

The detailed work of Zhmoginov et al. [32] (see also [33]) has informed the design of a new, vertically orientated apparatus named ALPHA-g [132], whose initial aim is to improve the limit of $|F|$ to around unity (i.e., a so-called "up-down" determination for the gravitational behaviour of antimatter). ALPHA-g consists of two symmetrically located atom and Penning trap arrangements at either end of the apparatus, with a high precision trapping region in the centre, with the magnetic potential in the vertical direction controlled by a series of coils. These coils will allow a bias field to be added to the bottom of the trap that can compensate for the gravitational potential difference across the trap. A determination of the gravitational acceleration of antihydrogen will be carried out by varying the compensation fields and monitoring the relative populations of the anti-atoms leaving the top and bottom of the trap [38]. A measurement accuracy of $|F| \sim 1$ can be approached using a few hundred trapped anti-atoms, as verified using the aforementioned simulations, and something that can be achieved in a single 8-hour antiproton shift at the AD. Bertsche [132] has also argued that with augmentation of ALPHA-g using various techniques such as in-situ magnetometry and antihydrogen laser cooling, it should be possible to lower the systematic errors on the measurement of $|F|$ to the 1% level.

An extension of capability for the vertical ALPHA-g has been suggested by Müller and co-workers [133] who envisage using a novel light-pulse (anti)matter-wave interferometer (see e.g. [134]) with trapped and cooled antihydrogen atoms, which are then released into the device. Without going into further details here, Müller et al. envisage a so-called basic scenario, capable of probing $|F|$ to around 1%, with further advances possibly allowing measurements approaching the ppm-level.

The AEgIS collaboration (e.g. [130]) is planning to use a beam of antihydrogen atoms formed via the reaction of very cold ($\sim$a few mK) antiprotons with excited positronium atoms (i.e. $\overline{p} + Ps^* \rightarrow \overline{H} + e^-$, a reaction first suggested as a useful source of $\overline{H}$ some time ago [135]) to perform a moiré deflectrometry experiment. The relative sensitivity to the gravitational acceleration g is expected to be around 1%, and a demonstration of the technique using a flux of fast $\overline{p}$s has recently been reported [136].

GBAR is planning to perform free fall experiments on ultra-cold antihydrogen atoms formed via photoionisation of cold antihydrogen positive ions, $\overline{H}^+$. The latter is to be produced following the $\overline{p}$–Ps* reaction (see above), and a further charge exchange as $\overline{H} + Ps \rightarrow \overline{H}^+ + e^-$, most probably with ground state Ps. The $\overline{H}^+$ will likely be formed at kinetic energies in the keV range, so will be decelerated and then individually sympathetically cooled in a Paul-type charged particle trap (using a laser-cooled Be^+ ion) into the mK regime. It will then be photo-ionised, allowing the resulting antihydrogen atom to undergo free fall: its time of flight between the pulsed laser used for ionisation and the subsequent annihilation on the chamber wall will be used to determine g.

A further free fall approach has been proposed by Voronin and co-workers [137, 138] which relies upon the interaction of cold antihydrogen with a material surface. Low energy antihydrogen will, due to quantum effects arising from the Casimir–Polder interaction, be efficiently quantum reflected from a surface and, in the presence of the Earth's gravitational field, will form long-lived quantum states. It has been shown [137, 138] that measuring the difference in the energy of the states using atom interferometry can yield a value for g for antihydrogen. With the anti-atoms at temperatures of $\sim 100\,$mK, it is claimed that a flux of a few antihydrogen atoms per second can yield a precision of around 10^{-3}.

Finally, in addition to these free-fall experiments, there is the possibility in future of performing a gravitational redshift experiment of the Pound–Rebka type directly on antihydrogen. Some basic theoretical considerations are discussed in Sect. 3.3.2.

3.3.1 Antihydrogen Free Fall

We now consider some of the theoretical ideas introduced in Sects. 2.5 and 2.6 on how GR may be modified, or extended, to predict violations of the weak equivalence principle (WEPff) in experiments on antimatter, specifically antihydrogen.

The difficulty of course is that existing experiments already place extremely small limits on any possible WEPff violation in matter systems. If there is to be any chance of measuring deviations from WEPff in the forthcoming antihydrogen free-fall experiments, we therefore need to find some mechanism which is effectively shielded in matter interactions but leaves a residual effect on antimatter.

Strong Equivalence Principle Violation

As described earlier, the simplest modification of GR is to break the strong equivalence principle (SEP) by introducing direct couplings of the elementary particle fields to the local curvature, for example generalising the Dirac action to:

$$S = \int d^4x \sqrt{-g} \left(\frac{R}{16\pi G} + \bar{\psi} \left(i\gamma^\mu D_\mu - m \right) \psi + a\, \partial_\mu R\, \bar{\psi}\gamma^\mu \psi + c\, R_{\mu\nu}\, \bar{\psi} i\gamma^\mu \overleftrightarrow{D}^\nu \psi + \cdots \right).$$

$$(3.43)$$

These extra terms (see (2.51)) may be viewed either as new couplings in a fundamental theory, or as an effective theory where they are generated by quantum loop corrections. In either case, they modify the fermion dispersion relation and the equation of motion for free-fall, which is then no longer the geodesic equation. Moreover, defined in a local inertial frame, the operator $\bar{\psi}\gamma^a\psi$ is CPT odd and its coupling to $\partial_\mu R$ modifies the geodesic equation *differently* for fermions and antifermions.

Unfortunately, it seems this mechanism cannot be exploited for free-fall experiments on Earth, since the Schwarzschild metric describing the gravitational field outside the source region is Ricci flat—only the Riemann curvature $R_{\mu\rho\nu\sigma}$ is nonzero, while $R_{\mu\nu}$ and R vanish. As noted in [67], it is not possible to construct a term bilinear in the Dirac fields in (3.43) involving $R_{\mu\rho\nu\sigma}$ which cannot be reduced at linear order in curvature to those expressible in terms of the Ricci tensor alone.

We conclude that in general relativity itself, even allowing for the SEP-violating interactions in the effective field theory, no observable WEPff violations are predicted in free-fall experiments in the Earth's gravitational field.

Lorentz and CPT *Violation*

A more radical alternative is to take the local Lorentz and CPT violating effective field theory discussed at length above and couple it to gravity. Incorporating spontaneous Lorentz violation into GR is not without subtlety, however, and the resulting theory involves many features requiring an extensive theoretical analysis [139–141]. Here, we just present a simplified account of how this could affect antihydrogen free-fall.

We consider for simplicity only two of the possible couplings, with the action:

$$S = \int d^4x \sqrt{-g}\left(\frac{R}{16\pi G} + \bar{\psi}\left(i\gamma^\mu D_\mu - m\right)\psi + a_\mu\,\bar{\psi}\gamma^\mu\psi + c_{\mu\nu}\,\bar{\psi}i\gamma^\mu D^\nu\psi + \cdots \right). \tag{3.44}$$

The analogy with (3.43) is obvious. Here, however, a^μ and $c_{\mu\nu}$ are entirely new background fields. If they take fixed background values (or VEVs if they are considered as quantum fields) this selects a preferred direction in the local orthonormal frame at each point in spacetime, thereby breaking local Lorentz invariance.

To find the classical, single-particle equation of motion originating from the theory (3.44), we follow the method described in Sect. 2.4 where it led to the geodesic equation. First, we write the single-particle action in curved spacetime including the two new background fields as,

$$S = -m \int d\lambda \left(\sqrt{\left(g_{\mu\nu} + c_{\mu\nu}\right)u^\mu u^\nu} + a_\mu u^\mu \right), \tag{3.45}$$

with $u^\mu = dx^\mu/d\lambda$. Here, $x^\mu(\lambda)$ is the particle trajectory, with λ an affine parameter which we could choose directly as proper time. To motivate this, note that the field $c_{\mu\nu}$ effectively enters (3.44) as a modification to the metric, while a^μ is analogous to an external electromagnetic field. It follows that we should not take a_μ of the 'pure gauge' form $\partial_\mu\phi$, otherwise it could be absorbed into a phase redefinition of the fermion field in (3.44).

The modified geodesic equation is found as before by extremising this action with respect to $x^\mu(\lambda)$. Under various technical assumptions, in particular that $c_{\mu\nu}$ and a^μ are slowly varying, and using the fact that we can take $c_{\mu\nu}$ to be traceless, a short calculation yields the following equation of motion in the weak-field, non-relativistic ($u^i \ll u^0$) limit,[10]

$$m_i \frac{d^2 x^i}{dt^2} + m_g\, \partial_i U = 0\,, \qquad (3.46)$$

where

$$m_i = m\left(1 + \frac{5}{6}c_{00}\right)$$
$$m_g = m\left(1 + \frac{1}{2}c_{00} + a^0\right). \qquad (3.47)$$

As in Sect. 2.4, we have used $\Gamma^i_{00} \simeq -\frac{1}{2}h_{00,i}$ with the gravitational potential $U = -\frac{1}{2}h_{00} = -GM/r$ for the Schwarzschild metric.

The key point here is that the new background field values c_{00} and a^0 appear with *different* coefficients in front of the "acceleration" and "gravity" terms in this modified geodesic equation. See the discussion following (2.34). In the simple weak-field, non-relativistic limit, these can be interpreted as modifications to (or better, definitions of) the "inertial mass" and "gravitational mass" respectively. This is clearly a violation of WEPff.

Moreover, for the corresponding antimatter test particle, we replace the C and CPT odd field a^0 in (3.47) by $-a^0$. This means we have a *different* ratio of m_g/m_i for matter and antimatter.

Now, in order to satisfy the extremely strong limits $\Delta g/g = m_g/m_i - 1 \lesssim 10^{-15}$ on WEPff for matter [142], which of course are possible because we can perform matter experiments with large test bodies, we have to assume a specific relation between c_{00} and a^0 in this model. Imposing $m_g = m_i$ for matter fixes $a^0 = c_{00}/3$, so then for antimatter $\Delta g/g \simeq -2c_{00}/3$.

This is reminiscent of the mechanism introduced for new 'fifth force' background fields in Sect. 2.6. There, we exploited the fact that a vector field couples oppositely to particles and their antiparticles to arrange a cancellation of the effect of new background fields on matter while leaving a residual unconstrained effect on antimatter. The idea is similar here.

Of course, this particular choice of parameters is otherwise unmotivated and requires fine-tuning to evade existing equivalence principle bounds. Moreover, going beyond the leading non-relativistic, Newtonian-like approximation (3.46) to the equation of motion will introduce corrections at least of $O(v^2/c^2)$, and the precise cancellation required to shield the effect of the new a^μ and $c_{\mu\nu}$ fields on matter cannot be maintained for the whole range of WEPff tests involving different velocities.

---

[10]An essentially identical equation follows from a careful treatment of the VEVs and fluctuations of $c_{\mu\nu}$ and a_μ in the framework of the gravitationally coupled SME, as detailed in [115, 139–141].

Depending on the test considered (compare Sect. 2.6), the size of potential WEPff violations in antimatter would then be limited to around $\Delta g/g \lesssim 10^{-7}$. Identifying c_{00} in (3.47) with the parameter bounded by the absence of observed Lorentz violation through annual variations in the hydrogen spectrum [109] would also limit $\Delta g/g$ for antihydrogen in this model.

New Background Fields

In Sect. 2.6, we discussed at some length the implications of new long-range scalar, vector or tensor (S, V, T) fields for WEPff violations. In particular, we discussed the conditions under which it may be possible to limit WEPff violations in matter experiments to satisfy the current experimental bounds while retaining a possible observable signal with antimatter. This discussion, and the associated bounds, apply directly to the forthcoming antihydrogen free-fall experiments.

Various other far more speculative models and suggestions have been advanced to try and justify an $O(1)$ WEPff violating effect with antimatter, none of which in our view is theoretically well-founded or can be made compatible with the huge body of experimental and observational evidence supporting GR and the standard model. Straightforward phenomenologies, such as the parametrisation $(g_{00})_{\text{eff}} = 1 - \alpha_g 2GM/r$ of the metric at the Earth's surface with α_g chosen independently for antimatter, give rise in the obvious way to a non-vanishing $\Delta g/g$, given by

$$\frac{\Delta g}{g} = \alpha_g - 1 . \tag{3.48}$$

However, as already noted in Sect. 2.5, this is unmotivated in terms of fundamental theory and apparently inconsistent since it would imply a direct dependence on the absolute gravitational potential and not, more realistically, only on a potential difference.[11] Even so, as we note below, redshift experiments involving antimatter interpreted with this parametrisation typically bound $|\alpha_g - 1| \lesssim 10^{-6}$. Certainly, there is no indication of the extreme "antigravity" value $\alpha_g = -1$.

The overall conclusion from theory is therefore that while possible violations of WEPff in antihydrogen free-fall can be envisaged, every viable model suggests that they are extremely small, almost certainly already constrained at the $\Delta g/g \lesssim 10^{-7}$ level. In this light, the current generation of antihydrogen gravity experiments should be regarded as an important first step, with the ultimate goal of reaching this realistic, but challenging, level of precision.

---

[11] A *caveat* here is that we can devise experiments analogous to the Bohm–Aharanov effect in gauge theories in which a redshift may be affected by the gravitational potential without a direct gravitational force acting [143]. However, as in the Bohm–Aharanov effect, this requires an element of non-trivial topology which is not the case for the simple free-fall experiments described here.

3.3.2 Antihydrogen Spectrum and Gravitational Redshift

Throughout the discussion so far, we have seen many ways in which the free-fall equivalence principle (WEPff) may be violated, even in ways which differ depending on whether the test particle is matter or antimatter.

On the other hand, none of these models has challenged the fundamental assumption of the 'universality of clocks' equivalence principle (WEPc), which asserts that all ideal clocks measure the same gravitational time dilation. This is ensured in general relativity by the fact that time measurements are determined by the metric component g_{00} and all matter (and antimatter) couples universally to the same metric.

Of course, the identification of an "ideal" clock is not straightforward and we have discussed how atomic spectral frequencies in atoms (and anti-atoms) may mimic WEPc violations due to unconventional new interactions or Lorentz or CPT violation. These effects must be identified and eliminated before we can attribute any anomalies to a genuine gravitational WEPc violation. For example, as described in the discussion near the end of Sect. 3.2.2, studying possible annual variations in the spectra of hydrogen and antihydrogen during the Earth's elliptical orbit round the Sun would allow both the SME Lorentz violating parameter c_{200}^{NR} and the Lorentz and CPT violating parameter a_{200}^{NR} to be determined separately, and isolated from any WEPc violating effect dependent on the gravitational field.

Theoretically, one way to achieve a genuine WEPc violation is to construct a multi-metric theory (see e.g. [144, 145]) in which different particle species impose different metric structures on the underlying spacetime manifold. A discussion of these theories is beyond the scope of this article but, even so, no causal QFT of this type has been established for which particles and their antiparticles couple to different metrics.

In this section, therefore, we consider tests of gravitational time dilation and redshift with antimatter, specifically antihydrogen, in the standard framework of GR. The essential results have already been presented in Sect. 2.4. To make contact with existing literature, we also comment on the phenomenological parametrisation of WEPc violations described, and criticised, at the end of Sect. 2.5.

To begin, we reflect on how in principle a frequency measurement of an atomic transition such as the $1S$–$2S$ antihydrogen line is made, taking GR into consideration. Essentially, the desired transition is induced using a laser with a tuned frequency which is in turn locked onto a reference time standard such as a co-located Cs atomic clock. The key point is that ultimately any frequency or time measurement is in fact simply a *ratio* of the frequency/time interval to be measured with another atomic transition frequency characterising the reference clock.

Now, as we have seen in Sect. 2.4, an atomic transition frequency in a gravitational field will be redshifted proportionally to the local gravitational potential. However, the *same* redshift applies to the co-located reference clock. So, to take a specific example, the $1S$–$2S$ antihydrogen frequency measured by a co-located (and co-moving) Cs atomic clock will remain the same, independent of the local gravitational potential. Colloquially, though imprecisely, we may say that time is running slowly in the gravitational field for both the measured atom and the reference clock, but crucially—according to GR—at the same rate.

Similarly, if we consider measurements taken through the annual cycle of the Earth's elliptical orbit of the Sun, so that the atom and reference clock are co-moving through the varying gravitational field of the Sun, the measured atomic transition frequency remains the same.

The gravitational redshift can however be detected if the atom and reference clock are not co-located, or not co-moving. In this case we would compare the atomic transition frequencies with the atom at different points with differing gravitational potentials as measured by a reference clock which remains fixed. The analysis of this situation in the idealised context of the Pound–Rebka experiment has been described in Sect. 2.4. Here, two atoms at different heights in the Earth's gravitational field compare frequencies through the direct exchange of a photon. In practice, this could be performed with an extended ALPHA-g apparatus with an upper and lower trap. The antihydrogen $1S$–$2S$ frequencies in these upper and lower traps would then be measured by lasers with frequencies calibrated to a single *fixed* reference Cs clock. This allows a measurement of the ratios of the frequencies of the upper and lower atoms for which the GR prediction is

$$\frac{\Delta \nu}{\nu} = -\frac{GM_E h}{R_E^2} . \tag{3.49}$$

Here, $\Delta \nu$ is the difference of the upper and lower frequencies, while the height difference is h. The key point of course is that this is dependent only on the *difference* of the gravitational potentials in the upper and lower trap. This gives

$$\frac{\Delta \nu}{\nu} = -1.1 \times 10^{-16} \frac{h}{1\,\text{m}} . \tag{3.50}$$

With the frequency precision currently attained with hydrogen, this gravitational redshift effect could in principle be measured with a height difference of order $h \simeq 10\,\text{m}$. Of course, a practical realisation of this experiment would nevertheless present many challenges. Note also that, according to GR, this redshift is universal and should be the same for antihydrogen and hydrogen, the matter-antimatter distinction being irrelevant.

Naturally, any deviations from these predictions would constitute an effective violation of WEPc and, if all non-gravitational origins could be excluded, would be in direct conflict with general relativity.

At this point, we would ideally have a well-defined extension of GR against which to compare any anomalous measurements and constrain new parameters. The gravitational extension of the Lorentz and CPT-violating SME is one such model and its implications for a variety of types of clock measurements have been extensively discussed in [116, 140, 141]. Without invoking Lorentz violation, a popular phenomenological parametrisation of possible beyond-GR effects was introduced by Hughes and Holzscheiter [70]. It essentially violates the weak equivalence principle by asserting that different particle species couple to different spacetime metrics, as described in Sect. 2.5. Equivalently, it conjectures that the coupling of matter to gravity, in the weak field approximation, is given by the Lagrangian,

$$L = \frac{1}{2} \alpha_g \, h_{\mu\nu} T^{\mu\nu} \,, \tag{3.51}$$

with the deviation from the flat spacetime metric given by $h_{\mu\nu}$ with $h_{00} = h_{rr} = -2U$ for the Schwarzschild metric, where $U = -GM/r$ is the local gravitational potential.

The violation of GR comes in allowing α_g to differ for different particle species, and for particles and their antiparticles, removing the universality of matter couplings to gravity embodied in the equivalence principle. It has to be immediately recognised, however, that (3.51) cannot be embedded in a relativistic QFT of the type used so successfully in the standard model, and indeed the Lorentz-violating SME. In these theories, based on causal fields lying in representations of the Lorentz group (see Sect. 2.1), the energy-momentum tensor is necessarily built from fields describing both particles and antiparticles.

Nevertheless, in the absence of an alternative well-motivated theory, we can still parametrise the scenarios described above in this model. The simplest case is to compare the antihydrogen and hydrogen $1S$–$2S$ transition frequency at the same place, measured through the intermediary of a reference Cs atomic clock. The key conceptual point here is that if the period of the antihydrogen transition being observed is dT in flat spacetime, then in a gravitational field this is the period *in the atom's proper time*, i.e. $d\tau = dT$, where $d\tau = \sqrt{-g_{00}(\alpha_g^{\bar{H}})} \, dt$ with $g_{00}(\alpha_g^{\bar{H}}) = -\left(1 - \alpha_g^{\bar{H}} 2GM/r\right)$. It follows that the period *in coordinate time* is $dt = dT/\sqrt{-g_{00}(\alpha_g^{\bar{H}})}$. This is measured by the Cs reference clock in terms of its own proper time, which in this model is determined by a metric with a potentially different coefficient α_g^{Cs}. So, in terms of the flat spacetime frequency $\nu = 1/dT$, the frequency of the antihydrogen transition *as measured by the* Cs *atomic clock*, satisfies

$$\frac{\nu^{\bar{H}}}{\nu} = \frac{\sqrt{-g_{00}\left(\alpha_g^{\bar{H}}\right)}}{\sqrt{-g_{00}\left(\alpha_g^{Cs}\right)}} \simeq 1 + \left(\alpha_g^{\bar{H}} - \alpha_g^{Cs}\right) U \,. \tag{3.52}$$

Note though that since we cannot have a physical measurement of the atomic frequency in the absence of any gravitational field, the flat spacetime frequency ν is at best a theoretical, not a measured, quantity.

A more physically direct result follows immediately by comparing the frequencies of hydrogen and antihydrogen *as measured by the same reference clock* in a gravitational field. Details of the clock cancel from this ratio and we are left with the conceptually clear prediction,

$$\frac{\Delta \nu^{\bar{H}-H}}{\nu^H} \equiv \frac{\nu^{\bar{H}} - \nu^H}{\nu^H} = \left(\alpha_g^{\bar{H}} - \alpha_g^H \right) U \, . \tag{3.53}$$

We have assumed here that the flat spacetime frequencies of $\bar{H}$ and H are the same so, as elsewhere in this section, this formula applies if the possibility of CPT violation has been excluded a priori. Then, unless the coupling α_g is the same for $\bar{H}$ and H, there will be a difference in the transition frequencies, allowing a bound to be placed on $\Delta\alpha_g^{\bar{H}-H}$. Here, however, we are confronted most starkly with a basic problem of this phenomenological model. By violating WEPc, this model predicts that a physical quantity is dependent on the local value of the gravitational potential itself, not a difference in potentials. So what value of the potential is relevant here? At first sight, one might suppose that we should use the Earth's gravitational potential at the surface, $|U| \simeq GM_E/R_E \simeq 7 \times 10^{-10}$. However, the Sun's potential is bigger, $|U| \simeq 1 \times 10^{-8}$. In fact, the potential becomes greater for more distant gravitational structures. For the galaxy, we may estimate $|U| \simeq 10^{-6}$ [146] with even higher values for the local galactic cluster. The limit placed on $\Delta\alpha_g^{\bar{H}-H}$ entirely depends on this choice, with the most stringent bound coming from the highest potential. If we take the current precision of antihydrogen spectroscopy, and in the absence of annual variations, we would find $\Delta\alpha_g^{\bar{H}-H} \lesssim 10^{-6}$. However, this has little credence and we would find an even smaller bound if, for example, we chose the gravitational potential of the Virgo cluster.

At first sight more reasonably, though disguising the same fundamental difficulty with the model, we could consider frequency measurements which depend only on differences of the gravitational potential. So next we consider the bounds on $\alpha_g^{\bar{H}}$ that would arise from the non-observation of annual variations in the antihydrogen spectrum during the Earth's orbit in the Sun's gravitational field.

In the same way, we find the difference in antihydrogen frequencies measured by the Cs reference clock at two different distances r_1 and r_2 from the Sun is

$$\frac{\Delta \nu^{\bar{H}}(r_1 | r_2)}{\nu^{\bar{H}}} \equiv \frac{\nu^{\bar{H}}(r_1) - \nu^{\bar{H}}(r_2)}{\nu^{\bar{H}}(r_2)} = \left(\alpha_g^{\bar{H}} - \alpha_g^{Cs} \right) \Delta U_S(r_1 | r_2) \, , \tag{3.54}$$

where $\Delta U_S(r_1 | r_2)$ is the difference in the Sun's potential at r_1 and r_2, and as usual we quote the result to $O(U)$ only. An annual variation would therefore indicate a difference between the parameters $\alpha_g^{\bar{H}}$ and α_g^{Cs} characterising the antimatter antihydrogen atom and the matter Cs atom respectively.

As before, we can cancel out the characteristics of the reference clock and compare the measured $\bar{H}$ and H frequencies directly. We then find,

$$\frac{\Delta \nu^{\bar{H}-H}(r_1 | r_2)}{\nu^H} \equiv \frac{\nu^{\bar{H}} - \nu^H}{\nu^H} \bigg|_{r_1} - \frac{\nu^{\bar{H}} - \nu^H}{\nu^H} \bigg|_{r_2} = \left(\alpha_g^{\bar{H}} - \alpha_g^H \right) \Delta U_S(r_1 | r_2) \, , \tag{3.55}$$

giving a very direct antimatter-matter comparison depending only on a difference of gravitational potentials.

To quantify this, at aphelion $r_1 = 1.52 \times 10^{11}$ m while at perihelion $r_2 = 1.47 \times 10^{11}$ m, so for the Earth's orbit we find,

$$\Delta U_S(r_1|r_2) \simeq \frac{GM_S}{r_2^2} (r_1 - r_2) = 3.3 \times 10^{-10} . \tag{3.56}$$

So with the current antihydrogen $1S$–$2S$ precision of 10^{-12}, the non-observation of annual variations in $\bar{\text{H}}$ and H would place a bound $\Delta\alpha_g^{\bar{\text{H}}-\text{H}} \lesssim 3 \times 10^{-3}$, while if the antihydrogen precision would match that currently available with hydrogen, this bound could be improved to $\Delta\alpha_g^{\bar{\text{H}}-\text{H}} \lesssim 10^{-6}$. A search for annual variations in the antihydrogen spectrum could, if this model is adopted, provide a very competitive test of WEPc.

The analysis of an antihydrogen Pound–Rebka experiment in this model is a straightforward extension of the usual GR treatment given in Sect. 2.4. From (2.46) and (3.49), the difference of the frequencies of the 'emitter' and 'observer' antihydrogen atoms is,

$$\frac{\Delta\nu^{\bar{\text{H}}}}{\nu^{\bar{\text{H}}}} \equiv \frac{\nu_O - \nu_E}{\nu_E} = -\alpha_g^{\bar{\text{H}}} \frac{GM_E h}{r_E^2} . \tag{3.57}$$

As noted above, measurement of this redshift factor is at the limit of what may be attained even if the antihydrogen $1S$–$2S$ precision could match that of hydrogen. So this experiment, while in principle an excellent test of WEPc, would only be sensitive to $O(1)$ deviations from the GR value $\alpha_g^{\bar{\text{H}}} = 1$.

Finally, the gravitational redshift measurement using atom matter-wave interferometry is modified in the same way in the phenomenological WEPc violating model. All that changes in (2.47)–(2.50) is the inclusion of the relevant α_g factor modifying the gravitational potential. In particular, the final formula for the phase shift (2.50) becomes

$$\Delta\phi_{redshift} = \alpha_g^{\bar{\text{H}}} \omega_C \int_0^T dt\, g\, \Delta r(t) , \tag{3.58}$$

for an antihydrogen interferometry experiment. Again, this allows a bound to be placed on the antimatter WEPc violation factor $|\alpha_g^{\bar{\text{H}}} - 1|$, and the proposal [133] suggests that precisions of order $|\alpha_g^{\bar{\text{H}}} - 1| \lesssim 10^{-6}$ are possible.[12]

$U(1)_{B-L}$ and Gravitational Redshift

In previous sections, we have considered the *direct* effects of a possible long-range $U(1)_{B-L}$ field sourced by the Earth. Here, we consider briefly the *indirect* effects

---

[12]In an experiment of this type with Cs atoms, the gravitational redshift is observed to agree with GR for trajectory separations of $O(0.1\text{mm})$, giving a bound $|\alpha_g^{Cs} - 1| < 7 \times 10^{-9}$. Assuming no WEPc violation, this experiment can also bound the local gravitational acceleration, finding $\Delta g/g < 3 \times 10^{-9}$. See [64, 65, 134, 147] for further details and references to the literature on atom matter-wave interferometry.

arising from the modifications to the Schwarzschild metric induced by the Earth's $U(1)_{B-L}$ charge. We focus on gravitational redshift.

The picture is straightforward. A massless $U(1)_{B-L}$ gauge field sourced by the large Q_{B-L} charge of the Earth is entirely analogous to the electromagnetic field around a charged object. Its gravitational effects are therefore described by the analogue of the Reissner–Nordström metric, usually used to describe the spacetime around a static, charged black hole. This metric is therefore:

$$ds^2 = -\left(1 - \frac{2GM}{r} + \alpha' \frac{GQ_{B-L}^2}{r^2}\right) dt^2$$

$$+ \left(1 - \frac{2GM}{r} + \alpha' \frac{GQ_{B-L}^2}{r^2}\right)^{-1} dr^2 + r^2 \sin^2(\theta) d\phi^2 , \qquad (3.59)$$

where $\alpha' = g'^2/4\pi$ is the fine structure constant corresponding to the fundamental $U(1)_{B-L}$ coupling, and Q_{B-L} is the numerical $U(1)_{B-L}$ charge of the source, of mass M.

The standard analysis of gravitational redshift now follows from our earlier discussion, with the new metric component

$$- g_{00}(r) = 1 - \frac{2GM}{r} + \alpha' \frac{GQ_{B-L}^2}{r^2} . \qquad (3.60)$$

Since this new metric is still a function of r only, the redshift derivation presented in Sect. 2.4 applies directly. With an emitter at radius r_E above the Earth's centre and a receiver at $r_O = r_E + h$, the ratio of frequencies measured at r_O and r_E is (see 2.46),

$$\frac{\nu_O}{\nu_E} = \frac{\sqrt{-g_{00}(r_E)}}{\sqrt{-g_{00}(r_O)}} , \qquad (3.61)$$

and we find the leading contributions,

$$\frac{\nu_O}{\nu_E} = 1 - \frac{GM_E}{R_E^2} h + \alpha' \frac{GQ_{B-L}^2}{R_E^3} h + \cdots \qquad (3.62)$$

with M_E, R_E the mass and radius of the Earth.

To establish the relative size of this new contribution to the redshift, which is of course the same for a matter or antimatter clock, we take the ratio of the final two terms in (3.62), which is approximately $10^{28}\alpha'$. Verification of the GR prediction for the redshift would therefore constrain $\alpha' \lesssim 10^{-28}$. Small though this is, it is still however many orders of magnitude bigger than the experimental limit already imposed by conventional equivalence principle tests (Sect. 2.6) which require $\alpha' \lesssim 10^{-49}$.

Alternatively, we could consider the redshift effect due to the eccentricity of the Earth's orbit around the sun, for which the corresponding ratio is still comparable, roughly $10^{30}\alpha'$. Despite the theoretical elegance of this theory, it therefore seems that there is no realistic possibility to detect a long-range $U(1)_{B-L}$ field even with the most sensitive gravitational redshift experiments.

Finally, we should note that the modification (3.60) of the metric also alters the equation of motion for free-fall, replacing the effective gravitational potential $U = -\frac{1}{2}h_{00} = -GM/r$ with

$$V = -\frac{GM}{r} + \frac{\alpha'}{2}\frac{GQ_{B-L}^2}{r^2} . \tag{3.63}$$

This induces an extra $1/r^3$ component in the gravitational force, though again the order of magnitude is far too small to be detectable given the existing constraints on the coupling α'.

Chapter 4
Other Antimatter Species

In this section, we discuss briefly some possibilities for testing fundamental physics principles using antimatter species other than antihydrogen. In Table 4.1 we have provided a summary of some of the antimatter species that have been, or may be in the not-too-distant future, subject to investigation and have indicated the types of test that may be performed, organised according to the discussion in Chap. 2. The types of experiments that permit such tests are also shown. Note that we confine ourselves to laboratory tests involving low ($\approx$eV or below) kinetic energies, and do not consider high energy measurements such as the special storage ring experiments aimed at the muon $g - 2$ value [148].

Most of the systems listed are stable against decay or self-annihilation. The notable exceptions are the bound-states muonium Mu ($\mu^+ e^-$) [149, 150], positronium Ps ($e^+ e^-$) [151, 152] and antiprotonic helium $He^+\bar{p}$, the metastable bound state of an antiproton and a helium ion [153, 154]). All these have, however, been studied spectroscopically and are the subject of ongoing investigations relevant to the types of fundamental physics tests described here.

As indicated in Table 4.1, the antiparticle species we consider are the positron, the antiproton and the antideuteron $\bar{d}$. Though heavier antibaryons have been created, e.g. anti-^{3}He [155] and the anti-alpha particle [156], their yields are currently too small to permit precision experimentation. The $\bar{d}$ has also not so far been subjected to such study, but it can be produced at $\approx 10^{-3}$ of the $\bar{p}$ flux and may be amenable to capture and manipulation. The possibility of doing was briefly discussed some time ago [157, 158]. The antineutron [159] is not amenable to capture and, like the neutron, is expected to be unstable as a free particle [93].

The antiproton and positron can be held for experimentation for extended periods (several months or longer, if required [160–162]) in charged particle traps. The latter are typically so-called Penning traps (or variants thereof) [163, 164] in which an harmonic electrical potential is used, together with a uniform magnetic field, to confine charged species. Measurement of the motion of the particles, and perhaps the spin-flip (with respect to the direction of the applied magnetic field), can determine properties such as the charge-to-mass ratio (often interpreted as a direct mass

© The Author(s), under exclusive license to Springer Nature Switzerland AG 2020
M. Charlton et al., *Antihydrogen and Fundamental Physics*,
SpringerBriefs in Physics, https://doi.org/10.1007/978-3-030-51713-7_4

Table 4.1 The antimatter particles and bound states discussed in this chapter, together with their electric charge and $B - L$ quantum number, the type of fundamental principles which they enable to be tested, and the types of experiments possible. WEPff and WEPc, as defined in Chap. 1, refer to the universality of free-fall and the universality of clocks respectively. AI denotes atomic matter-wave interferometry. We have only shown this in the table for the neutral antihydrogen, although AI experiments with other species may also be feasible

Species	$Q, B - L$	Tests	Experiments
$\overline{p}$	$-1, -1$	CPT, WEPc, Lorentz	Traps
$\overline{d}$	$-1, -2$	CPT, WEPc, Lorentz	Traps
e^+	$1, 1$	CPT, WEPc, Lorentz	Traps
$\overline{H}$	$0, 0$	CPT, WEPc, WEPff, Lorentz	Spectroscopy, AI, free fall
$\overline{D}$	$0, -1$	CPT, WEPc, WEPff, Lorentz	Spectroscopy, free fall
$\overline{H}^+$	$1, 1$	CPT, WEPc, Lorentz	Traps
$\overline{H}_2^-$	$-1, -1$	CPT, WEPc, Lorentz	Traps, Spectroscopy
Mu	$0, 0$	WEPc, WEPff, Lorentz	Spectroscopy, free fall
Ps	$0, 0$	WEPc, WEPff, Lorentz	Spectroscopy, free fall
$He^+\overline{p}$	$0, 2$	CPT, WEPc, Lorentz	Spectroscopy

measurement under the assumption that the fundamental charge is quantised and is equal and opposite for particles and antiparticles) and the $g - 2$ ratio.

Recent highlights have included the work of the BASE collaboration that has provided systematic improvements in $\overline{p}$ storage, manipulation and interrogation to yield values of the charge-to-mass ratio and magnetic moment to unprecedented accuracies [125, 126, 165]. For example, in [165], the ratio of the charge-to-mass ratios for the antiproton and proton was determined through cyclotron frequency comparisons as

$$\frac{q/m\,(\overline{p})}{q/m\,(p)} - 1 \; < \; 10^{-12} \,, \tag{4.1}$$

improving on a previous precision of $< 9 \times 10^{-11}$ [166]. While loosely interpreted as a high-precision test of CPT, as we have discussed in Chap. 2 this is really a test of more fundamental principles, particularly causality. The magnetic moment measurements in [125, 126] can be interpreted in the SME as a bound on the coefficient $|b_3^p| \lesssim 10^{-24}$ GeV [44], and also place stringent bounds on Lorentz violation through the absence of sidereal variations. On the other hand, if we assume no other non-standard physics, (4.1) could be viewed as a WEPc equivalence principle test. Interpreted in terms of the phenomenological model (3.51), and using the local galaxy supercluster gravitational potential $U \simeq 3 \times 10^{-5}$, the equality of the charge-to-mass ratios places a bound of $|\alpha_g^{\bar{p}} - 1| < 8.7 \times 10^{-7}$ (where again we assume $\alpha_g^p = 1$) [165].

Antihydrogen is of course the archetypal neutral antimatter bound state and is capable of investigation both via spectroscopy and in free fall. As described throughout this book, this offers the means of testing CPT and Lorentz symmetries with high precision as well as exploring the gravitational properties of antimatter with WEPc and WEPff tests.

Anti-deuterium ($\overline{\text{D}}$) should also become available for experiment in future, and offers further opportunities for complementary tests [38]. In terms of Lorentz and CPT violation as parametrised by the SME, its spectrum would be sensitive to SME couplings involving the antineutron as well as the antiproton [115], with the corresponding additions to the transition frequency calculations in Chap. 3. In principle, since unlike antihydrogen it has a non-zero $B - L$ charge, it would experience a violation in WEPff if there were a long-range $B - L$ interaction with the Earth. However, as with any antimatter species carrying a $B - L$ charge, since it necessarily follows that the corresponding matter system also has non-vanishing $B - L$, such interactions are already constrained to greater precision from studies of the equivalence principle with bulk matter.

Other current experimental work at CERN includes antiprotonic helium, $\text{He}^+\bar{p}$ [167]. This unique bound state is formed by stopping $\bar{p}$s in dilute He gas, with around 3% of the states formed being metastable against annihilation with lifetimes in the μs range. It has been the subject of a sustained programme of spectroscopic investigations (see e.g. [153, 154] for reviews). Recent highlights have included quasi-two-photon spectroscopy [107] and buffer gas cooling of the $\text{He}^+\bar{p}$ [168], which have allowed, for instance, the determination of some $\bar{p}$ properties to high precision. Possible types of fundamental physics test with $\text{He}^+\bar{p}$ are given in Table 4.1. Note that since there is no matter counterpart of $\text{He}^+\bar{p}$ available for experimental study, such tests require comparisons with detailed few body atomic structure calculations (see e.g. [169]). Laser spectroscopy of pionic helium πHe^+ has also been proposed and could allow a measurement of the π^- mass with a fractional precision 10^{-6}–10^{-8} [170].

The antihydrogen positive ion $\overline{\text{H}}^+$ ($\bar{p}\,e^+\,e^+$) and the antihydrogen molecular anion $\overline{\text{H}}_2^-$ ($\bar{p}\,\bar{p}\,e^+$) offer exciting possibilities for new tests of fundamental physics. These states have yet to be observed in the laboratory, but both have well studied and important matter counterparts and may be produced using interactions of trapped antihydrogen, or beams of the anti-atom. Possible mechanisms for $\overline{\text{H}}^+$ include radiative $\overline{\text{H}}/e^+$ combination and charge exchange in $\overline{\text{H}}/\text{Ps}$ collisions [171–173] and for $\overline{\text{H}}_2^-$, radiative $\overline{\text{H}}/\bar{p}$ association and $\overline{\text{H}}$-$\overline{\text{H}}$ associative attachment [174, 175]. The production of $\overline{\text{H}}^+$ is envisaged within the GBAR programme [173, 176] currently underway at the AD at CERN.

$\overline{\text{H}}^+$ is expected, as is its matter counterpart H^-, to be bound by around 0.75 eV and to have only a single bound state. As such, it is not amenable to spectroscopic investigation although it should be able to be stored for long periods in Penning traps for interrogation. $\overline{\text{H}}_2^-$ is a bound state expected to have a rich vibrational and rotational spectrum and H_2^+ has already been suggested as a possible ultra-precise optical clock [177]. Myers [174] has described how measurements of analogous clock transitions with $\overline{H}_2^-$ may be used as CPT tests with the potential for greater precision than is possible with antihydrogen. This is due primarily to the very long lifetimes (of the order of years, rather than seconds for antihydrogen) of some of the excited states.

Chapter 5
Summary and Outlook

In the last three years, the long-standing ALPHA programme of producing and trapping cold antihydrogen atoms in sufficient numbers and with the required control to permit precision spectroscopy has been achieved. This enables the anti-atom to be used for state-of-the-art precision tests with antimatter of the most fundamental principles of modern theories of particle physics and gravity, notably local Lorentz and CPT invariance, causality and, eventually, the various forms of the Equivalence Principle.

In Chap. 1 we briefly summarised how the development of sources of low (eV) energy positrons and antiprotons, and schemes for the accumulation and manipulation of the antiparticles culminated in the controlled creation of cold antihydrogen at the AD [178, 179]. Extending the techniques that made this possible, and superimposing neutral atom trapping technology in the form of a magnetic minimum atom trap onto the antihydrogen creation region, allowed the trapping of the anti-atom to be achieved some years later [21, 22].

This advance was key, particularly as long confinement times were quickly forthcoming [27], ensuring that the antihydrogen (initially produced in highly excited states, as discussed in Chap. 1) would decay to the ground state in readiness for experimentation. More recently, lifetimes in the trap in excess of 60 h have been achieved, allowing of the order of a thousand antihydrogen atoms to be accumulated on a shot-by-shot basis, thereby ensuring optimum use of the antiproton flux from the AD.

We have also summarised the main achievements in physics with antihydrogen, principally the work of the ALPHA collaboration. This has included: the setting of a limit on the charge neutrality of antihydrogen; the development of a method of investigating the behaviour of the anti-atom in the earth's gravitational field and the observation and characterisation of hyperfine and positronic transitions, including the seminal observation of the $1S–2S$ transition and a first determination of the Lamb shift. The pace of progress, with the $1S–2S$ transition already known to a few parts in 10^{12}, bodes well for future improvements in precision and the concomitant limits set upon fundamental physics principles.

© The Author(s), under exclusive license to Springer Nature Switzerland AG 2020   87
M. Charlton et al., *Antihydrogen and Fundamental Physics*,
SpringerBriefs in Physics, https://doi.org/10.1007/978-3-030-51713-7_5

On the theory side, we have discussed how each of these experiments tests the fundamental principles on which our current understanding of particle physics is based. We highlighted how these principles are interwoven in the structure of relativistic quantum field theories and their extension to include gravity through general relativity.

Particular emphasis was placed on the role of causality, and we reviewed how the existence of antiparticles with properties *exactly* matching those of the corresponding particles is required in a Lorentz invariant theory to preserve causality. The quantum field theory is then built from causal fields transforming in representations of the Lorentz group; these causal fields necessarily contain particles *and* antiparticles. A key feature of such local field theories is that they exhibit CPT symmetry—breaking CPT necessarily implies breaking Lorentz invariance and would undermine the fundamental principles on which our theories of particle physics are built, unlike the individual symmetries C, P, T, CP etc. which may be trivially broken in the standard model.

Gravity is introduced classically through GR; for the atomic physics experiments described here, the scales and weak gravitational fields involved mean that quantum gravity is irrelevant. The key feature of general relativity is that it describes gravitational forces by formulating the theory on a Riemannian curved spacetime. This is locally flat and it is in this local tangent space that relativistic QFT is formulated (so for example CPT symmetry has nothing to do with the curvature of the spacetime). Insisting on the strong equivalence principle, as defined here, imposes the requirement that the local causal fields couple only to the spacetime metric connection, *not* the curvature. All this realises a universality of matter-gravity couplings which implies the WEPff and WEPc equivalence principles.

This raises the question of how we could imagine breaking Lorentz or CPT symmetry or the EEP. Two minimal extensions of the standard model were discussed. First, an effective field theory in which additional local but non-Lorentz invariant (and possibly CPT odd) operators are added to the standard model was described. The new couplings in this SME model [39, 40, 115] may be included in calculations of atomic physics properties, especially in spectroscopy but also in modifying 'gravitational' and 'inertial' masses in free-fall experiments, and bounds placed on them. Here, we have carried this out explicitly for the frequencies associated with the particular transitions measured by ALPHA with the magnetic field in their confining trap. This allows quantitative bounds to be placed on Lorentz and CPT breaking and compared between quite different experiments.

Second, an effective field theory in which additional local operators coupling directly to the spacetime curvature [67] was described. This implies that in a local inertial frame, physical measurements are sensitive to the gravitational field strength in violation of SEP, and also WEPff. However, although this has interesting applications in cosmological spacetimes where it can give rise to matter-antimatter asymmetry through gravitational leptogenesis [68, 69], it was pointed out in Sect. 2.5 that this would not affect atomic physics experiments in the Earth's gravity. This is because above the Earth's surface, the gravitational field is described by a Ricci-flat spacetime and at leading order there are no fermion bilinear couplings to the Riemann tensor itself.

It was also emphasised that these theories must be regarded as low-energy *effective* field theories only. Although they break Lorentz invariance or SEP only minimally and are still built from local causal Lorentz fields, they do in general break causality. This was discussed in some detail here in Sect. 2.3. Causality is, however, a property of high-momentum propagation and it is possible for a theory which would be non-causal in itself to be the low-energy limit of a causal theory valid also at high energies. The existence of a causal UV completion (see Sect. 2.3) can impose special relations amongst the couplings in its low-energy approximation. Here, in the absence of any knowledge of this UV theory (whether a Planck-scale QFT, string theory or other formulation of quantum gravity) we have to consider the SME or SEP-violating couplings as arbitrary parameters to be fixed by experiment.

We also followed the literature by analysing certain equivalence principle experiments in terms of an entirely phenomenological model [70] in which matter and anti-matter were supposed to couple to different metrics, immediately violating WEPc and WEPff. This model was criticised, however, on two counts—first that it cannot be realised in terms of a quantum field theory with causal fields, and therefore cannot be related in a plausible way to the standard model, and second that they have the apparently unphysical feature of making local frequency comparisons dependent on the absolute value of the local gravitational potential, rather than on potential differences or the field strength.

The introduction of new 'fifth-force' interactions can also produce physical effects which mimic a genuine gravitational violation of the equivalence principle, SEP and WEPff in particular. We considered several potential new interactions, especially $U(1)_{B-L}$ gauge theories, supergravity inspired models and general features of gravitational-strength theories with new scalar, vector or tensor (S, V, T) fields. Of these, only vector-mediated interactions act with opposite sign on matter and antimatter (analogously to electromagnetism). A tensor interaction such as conventional gravity acts attractively on both matter and antimatter, coupling through the energy-momentum tensor and not some 'gravitational charge'. In early work, it was speculated that this could be exploited to hide WEPff violations for matter, for example by arranging a cancellation between new V and S forces for matter, while they would appear with double strength for antimatter where the sign of the V interaction would be reversed. We analysed these ideas in Sect. 2.5 with the conclusion that, partly due to the different velocity-dependence of S, V, T interactions, such cancellations cannot be exact. Existing bulk matter experiments therefore already place severe constraints on possible WEPff violations with antimatter systems, indeed several orders of magnitude below those accessible to the first-generation direct WEPff tests planned with antihydrogen.

This work began by looking at the ways in which individual experiments on antimatter, especially antihydrogen, test specific fundamental principles. What soon becomes apparent, however, is that it is generally impossible to associate a particular experiment with an unambiguous test of a particular principle, such as WEPc or CPT. In terms of theory, these fundamental properties are all part of a single, tight theoretical structure and individual elements cannot be violated without impacting the whole construction. The most obvious example is the CPT theorem, where with our current understanding of local relativistic QFT, CPT violation necessarily involves breaking Lorentz symmetry.

Experimentally, we saw how, for example, each transition frequency in the anti-hydrogen spectrum depends in a different way on the many couplings in the SME, each of which could be the place where Lorentz or CPT violation is hiding. Experiments such as the search for annual variations in the antihydrogen spectrum could be an indication either of Lorentz violation or WEPc violation, or indeed both. The possible existence of new fifth forces violating WEPff but not WEPc means that a null experimental finding limiting WEPc violation, such as the equality of q/m for protons and antiprotons, does not necessarily imply a null result in an antihydrogen free-fall experiment looking for WEPff violations.

In general, the moral of this is that in searching for presumably tiny violations of the fundamental principles of particle physics, *all* possible high precision experiments are valuable and well-motivated. In the event of an unexpected result in anti-matter physics, many complementary measurements would probably be required to pin down its implications for fundamental theory.

The outlook for experimental antihydrogen physics is currently excellent, and we can look forward to many new endeavours and potentially dramatic improvements of the precision of spectroscopic measurements. It is hoped that the new Extra Low ENergy Antiproton facility (ELENA) [29] will soon be fully operational at the AD. This should allow low energy antiproton capture efficiencies to increase by up to two orders of magnitude. Coupled with delivery to experiments using easily switchable electrostatic beamlines, this will result in much optimised use of the antiproton flux. One can envisage antihydrogen experiments working continuously, with much shorter frequency scan times, for instance, leading to much enhanced capabilities for the exploration and understanding of systematics and to measurement campaigns over extended periods of time. This promises a bright future for high precision tests of fundamental physics with antihydrogen.

References

1. Ahmadi, M., et al.: ALPHA collaboration. Nature **541**, 506 (2017)
2. Ahmadi, M., et al.: ALPHA collaboration. Nature **557**, 71 (2018)
3. Pauli, W.: Phys. Rev. **58**, 716 (1940)
4. Bell, J.S.: Birmingham University thesis (1954)
5. Lüders, G.: Det. Kong. Danske Videnskabernes Selskab Mat. fysiske Meddelelser **28**(5) (1954)
6. Pauli, W. (ed.): Niels Bohr and the Development of Physics. McGraw-Hill, New York (1955)
7. Damour, T.: Class. Quant. Grav. **29**, 184001 (2012). arXiv:1202.6311 [gr-qc]
8. Will, C.M.: Living Rev. Rel. **17**, 4 (2014). arXiv:1403.7377 [gr-qc]
9. Charlton, M., Humberston, J.W.: Positron Physics. Cambridge University Press, Cambridge (2000)
10. Schultz, P.J., Lynn, K.G.: Rev. Mod. Phys. **60**, 701 (1988)
11. Jørgensen, L.V., et al.: ATHENA collaboration. Phys. Rev. Lett. **95**, 025002 (2005)
12. Maury, S.: Hyperfine Interact. **109**, 43 (1997)
13. Eriksson, T.: Hyperfine Interact. **194**, 12 (2009)
14. Gabrielse, G., Fei, X., Helmerson, K., Rolston, S.L., Tjoekler, R.L., Trainor, T.A., Kalinowsky, H., Haas, J., Kells, W.: Phys. Rev. Lett. **57**, 2504 (1986)
15. Gabrielse, G., Fei, X., Orozco, L., Tjoekler, R.L., Haas, J., Kalinowsky, H., Trainor, T.A., Kells, W.: Phys. Rev. Lett. **89**, 1360 (1989)
16. Feng, X., Holzscheiter, M.H., Charlton, M., Hangst, J., King, N.S.P., Lewis, R.A., Rochet, J., Yamazaki, Y.: Hyperfine Interact. **109**, 145 (1997)
17. Bertsche, W.A., Butler, E., Charlton, M., Madsen, N.: J. Phys. B: At. Mol. Opt. Phys. **48**, 232001 (2015)
18. Ahmadi, M., et al.: ALPHA collaboration. Nature Commun. **8**, 681 (2017)
19. Holzscheiter, M.H., Charlton, M., Nieto, M.M.: Phys. Rep. **402**, 1 (2004)
20. Robicheaux, F.: J. Phys. B: At. Mol. Opt. Phys. **41**, 192001 (2008)
21. Andresen, G.B., et al.: ALPHA collaboration. Nature **468**, 673 (2010)
22. Gabrielse, G., et al.: ATRAP collaboration. Phys. Rev. Lett. **108**, 113002 (2012)
23. Foot, C.J.: Atomic Physics. Oxford Master Series in Atomic, Optical and Laser Physics. Oxford University Press, Oxford (2005)
24. Amole, C., et al.: ALPHA collaboration. Nucl. Inst. Meth. A **735**, 319 (2014)
25. Amole, C., et al.: ALPHA collaboration. Nucl. Inst. Meth. A **732**, 134 (2013)
26. Andresen, G.B., et al.: ALPHA collaboration. Phys. Lett. B **695**, 95 (2011)

© The Author(s), under exclusive license to Springer Nature Switzerland AG 2020
M. Charlton et al., *Antihydrogen and Fundamental Physics*,
SpringerBriefs in Physics, https://doi.org/10.1007/978-3-030-51713-7

27. Andresen, G.B., et al.: ALPHA collaboration. Nature Phys. **7**, 55 (2011)
28. Ahmadi, M., et al.: ALPHA collaboration. Nature **561**, 211 (2018)
29. Maury, S., et al.: Hyperfine Interact. **229**, 105 (2014)
30. Amole, C., et al.: ALPHA collaboration. Nature Commun. **5**, 3955 (2014)
31. Ahmadi, M., et al.: ALPHA collaboration. Nature **529**, 373 (2016)
32. Zhmoginov, A.I., Charman, A.E., Shalloo, R., Fajans, J., Wurtele, J.S.: Class. Quantum Grav. **30**, 205014 (2013)
33. Zhong, M., Fajans, J., Zukor, A.F.: New J. Phys. **20**, 053003 (2018)
34. Amole, C., et al.: ALPHA collaboration. Nature Commun. **4**, 1785 (2013)
35. Amole, C., et al.: ALPHA collaboration. Nature **483**, 439 (2012)
36. Ahmadi, M., et al.: ALPHA collaboration. Nature **548**, 66 (2017)
37. Parthey, C.G., et al.: Phys. Rev. Lett. **107**, 203001 (2011). arXiv:1107.3101 [physics.atom-ph]
38. The ALPHA Collaboration: "*A proposal for a project at the CERN Antiproton Decelerator after LS2 – Spectroscopic and gravitational measurements on antihydrogen: ALPHA-3, ALPHA-g and beyond*", CERN-SPSC-2019-036/SPSC-P-362 (2019)
39. Colladay, D., Kostelecký, V.A.: Phys. Rev. D **55**, 6760 (1997). arXiv:hep-ph/9703464
40. Colladay, D., Kostelecký, V.A.: Phys. Rev. D **58**, 116002 (1998). arXiv:hep-ph/9809521
41. Peskin, M.E., Schroeder, D.V.: An Introduction to quantum field theory. CRC Press, Boca Raton (1995)
42. Weinberg, S.: The Quantum Theory of Fields. Foundations, vol. 1. Cambridge University Press, Cambridge (2005)
43. McDonald, J.I., Shore, G.M.: JHEP **1604**, 030 (2016). arXiv:1512.02238 [hep-ph]
44. Kostelecký, V.A., Russell, N.: Rev. Mod. Phys. **83**, 11 (2011). arXiv:0801.0287v11 [hep-ph]
45. Klinkhamer, F.R.: Nucl. Phys. B **578**, 277 (2000). arXiv:hep-th/9912169
46. Klinkhamer, F.R.: J. Phys. Conf. Ser. **952**(1), 012003 (2018). arXiv:1709.01004 [hep-th]
47. M. Chaichian, A. D. Dolgov, V. A. Novikov, A. Tureanu, CPT violation does not lead to violation of Lorentz invariance and vice versa. Phys. Lett. B **699**, 177 (2011)
48. M. Chaichian, K. Fujikawa, A. Tureanu, Electromagnetic interaction in theory with Lorentz invariant CPT violation. Phys. Lett. B **718**, 1500 (2013)
49. K. Fujikawa, A. Tureanu, Lorentz invariant CPT breaking in the Dirac equation. Int. J. Mod. Phys. A **32**, 1741014 (2017)
50. M. Chaichian, K. Fujikawa, A. Tureanu, Lorentz invariant CPT violation. Eur. Phys. J. C **73**, 2349 (2013)
51. Greenberg, O.W.: Phys. Rev. Lett. **89**, 231602 (2002). arXiv:hep-ph/0201258
52. Kostelecký, V.A., Lehnert, R.: Phys. Rev. D **63**, 065008 (2001). arXiv:hep-th/0012060
53. Shore, G.M.: Contemp. Phys. **44**, 503 (2003). arXiv:gr-qc/0304059
54. Shore, G.M.: Nucl. Phys. B **778**, 219 (2007). arXiv:hep-th/0701185
55. Hollowood, T.J., Shore, G.M.: Nucl. Phys. B **795**, 138 (2008). arXiv:0707.2303 [hep-th]
56. Hollowood, T.J., Shore, G.M.: JHEP **0812**, 091 (2008). arXiv:0806.1019 [hep-th]
57. Hollowood, T.J., Shore, G.M., Stanley, R.J.: JHEP **0908**, 089 (2009). arXiv:0905.0771 [hep-th]
58. Hollowood, T.J., Shore, G.M.: JHEP **1202**, 120 (2012). arXiv:1111.3174 [hep-th]
59. McDonald, J.I., Shore, G.M.: JHEP **1502**, 076 (2015). arXiv:1411.3669 [hep-th]
60. Abbott, B.P., et al.: LIGO scientific and virgo collaborations. Phys. Rev. Lett. **116**, 061102 (2016). arXiv:1602.03837 [gr-qc]
61. Hobson, M.P., Efstathiou, G.P., Lasenby, A.N.: General Relativity: An Introduction for Physicists. Cambridge University Press, Cambridge (2006)
62. Visser, M.: Post-Newtonian particle physics in curved spacetime. arXiv:1802.00651 [hep-ph]
63. Pound, R.V., Rebka, G.A.: Phys. Rev. Lett. **4**, 337 (1960)
64. Müller, H., Peters, A., Chu, S.: Nature **463**, 926 (2010)
65. Hohensee, M.A., Chu, S., Peters, A., Müller, H.: Phys. Rev. Lett. **106**, 151102 (2011). arXiv:1102.4362 [gr-qc]
66. Blas, D.: Phil. Trans. Roy. Soc. Lond. A **376**, 20170277 (2018)
67. Shore, G.M.: Nucl. Phys. B **717**, 86 (2005). [hep-th/0409125]

68. McDonald, J.I., Shore, G.M.: Phys. Lett. B **751**, 469 (2015). arXiv:1508.04119 [hep-ph]
69. McDonald, J.I., Shore, G.M.: Phys. Lett. B **766**, 162 (2017). arXiv:1604.08213 [hep-ph]
70. Hughes, R.J., Holzscheiter, M.H.: Phys. Rev. Lett. **66**, 854 (1991)
71. Heeck, J.: Phys. Lett. B **739**, 256 (2014). arXiv:1408.6845 [hep-ph]
72. Ruegg, H., Ruiz-Altaba, M.: Int. J. Mod. Phys. A **19**, 3265 (2004). arXiv:hep-th/0304245
73. Basso, L., Belyaev, A., Moretti, S., Shepherd-Themistocleous, C.H.: Phys. Rev. D **80**, 055030 (2009). arXiv:0812.4313 [hep-ph]
74. Zachos, C.K.: Phys. Lett. B **76**, 329 (1978)
75. Zachos, C.K.: UMI-79-19867. https://thesis.library.caltech.edu/4718/1/Zachos_ck_1979.pdf
76. Scherk, J.: Phys. Lett. B **88**, 265 (1979)
77. Scherk, J.: Gravitation at Short Range and Supergravity. Talk presented at the Europhysics Study Conference "Unification of the Fundamental Interactions", Erice (Italy), March 1980; LPTENS-80-15, C80-03-17-26
78. Bellucci, S., Faraoni, V.: Phys. Lett. B **377**, 55 (1996). arXiv:hep-ph/9605443
79. Brans, C., Dicke, R.H.: Phys. Rev. **124**, 925 (1961)
80. Horndeski, G.W.: Int. J. Theor. Phys. **10**, 363 (1974)
81. Bekenstein, J.D.: Phys. Rev. D **70**, 083509 (2004); Erratum: Phys. Rev. D **71**, 069901 (2005). arXiv:astro-ph/0403694
82. Moffat, J.W.: JCAP **0603**, 004 (2006). [gr-qc/0506021]
83. Clifton, T., Ferreira, P.G., Padilla, A., Skordis, C.: Phys. Rep. **513**, 1 (2012). arXiv:1106.2476 [astro-ph.CO]
84. Macrae, K.I., Riegert, R.J.: Nucl. Phys. B **244**, 513 (1984)
85. Goldman, T.J., Hughes, R.J., Nieto, M.M.: Phys. Lett. B **171**, 217 (1986)
86. M. M. Nieto, T. J. Goldman and R. J. Hughes, *Phys. Rev. D* **36** (1987) 3684, *Phys. Rev. D* **36**, 3688 (1987); *Phys. Rev. D* **36**, 3694 (1987); *Phys. Rev. D* **38**, 2937 (1988)
87. Nieto, M.M., Goldman, T.J.: Phys. Rep. **205**, 221 (1991)
88. Williams, J.G., Turyshev, S.G., Boggs, D.: Class. Quant. Grav. **29**, 184004 (2012). arXiv:1203.2150 [gr-qc]
89. Adelberger, E.G., Heckel, B.R., Stubbs, C.W., Su, Y.: Phys. Rev. Lett. **66**, 850 (1991)
90. Alves, D.S.M., Jankowiak, M., Saraswat, P.: (2009). arXiv:0907.4110 [hep-ph]
91. Wagner, T.A., Schlamminger, S., Gundlach, J.H., Adelberger, E.G.: Class. Quant. Grav. **29**, 184002 (2012). arXiv:1207.2442 [gr-qc]
92. Aghanim, N., et al.: [Planck Collaboration], arXiv:1807.06209 [astro-ph.CO]
93. Tanabashi, M., et al.: Particle Data Group. Phys. Rev. D **98**, 030001 (2018)
94. Sakharov, A.D.: Pisma Zh. Eksp. Teor. Fiz. **5**, 32 (1967); JETP Lett. **5**, 24 (1967); Sov. Phys. Usp. **34**(5), 392 (1991); Usp. Fiz. Nauk **161**(5), 61 (1991)
95. Klinkhamer, F.R., Manton, N.S.: Phys. Rev. D **30**, 2212 (1984)
96. Kuzmin, V.A., Rubakov, V.A., Shaposhnikov, M.E.: Phys. Lett. B **155**, 36 (1985)
97. Buchmuller, W., Peccei, R.D., Yanagida, T.: Ann. Rev. Nucl. Part. Sci. **55**, 311 (2005). arXiv:hep-ph/0502169
98. Shaposhnikov, M.: J. Phys. Conf. Ser. **171**, 012005 (2009)
99. Garbrecht, B.: Why is there more matter than antimatter? Calculational methods for leptogenesis and electroweak baryogenesis. arXiv:1812.02651 [hep-ph]
100. Cohen, A.G., Kaplan, D.B.: Phys. Lett. B **199**, 251 (1987)
101. Kolb, E.W., Turner, M.S.: Front. Phys. **69**, 1 (1990)
102. Bertolami, O., Colladay, D., Kostelecký, V.A., Potting, R.: Phys. Lett. B **395**, 178 (1997). [hep-ph/9612437]
103. Kostelecký, V.A., Mewes, M.: Phys. Rev. D **88**, 096006 (2013). arXiv:1308.4973 [hep-ph]
104. Konopka, T.J., Major, S.A.: New J. Phys. **4**, 57 (2002). arXiv:hep-ph/0201184
105. Mavromatos, N.E., Sarkar, S.: Universe **5**, 5 (2018). arXiv:1812.00504 [hep-ph]
106. Ahmadi, M., et al.: ALPHA collaboration. Nature **578**, 375 (2020)
107. Hori, M., et al.: Nature **475**, 484 (2011)
108. Hughes, R.J., Deutch, B.I.: Phys. Rev. Lett. **69**, 578 (1992)
109. Matveev, A., et al.: Phys. Rev. Lett. **110**, 230801 (2013)

110. Bluhm, R., Kostelecký, V.A., Russell, N.: Phys. Rev. Lett. **82**, 2254 (1999)
111. Kostelecký, V.A., Lane, C.D.: Phys. Rev. D **60**, 116010 (1999). arXiv:hep-ph/9908504
112. Kostelecký, V.A., Lane, C.D.: J. Math. Phys. **40**, 6245 (1999). arXiv:hep-ph/9909542
113. Altschul, B.: Phys. Rev. D **81**, 041701 (2010). arXiv:0912.0530 [hep-ph]
114. Yoder, T.J., Adkins, G.S.: Phys. Rev. D **86**, 116005 (2012). arXiv:1211.3018 [hep-ph]
115. Kostelecký, V.A., Vargas, A.J.: Phys. Rev. D **92**, 056002 (2015). arXiv:1506.01706 [hep-ph]
116. Kostelecký, V.A., Vargas, A.J.: Phys. Rev. D **98**, 036003 (2018). arXiv:1805.04499 [hep-ph]
117. Caswell, W.E., Lepage, G.P.: Phys. Lett. B **167**, 437 (1986)
118. Labelle, P.: Phys. Rev. D **58**, 093013 (1998). arXiv:hep-ph/9608491
119. Gomes, Y.M.P., Malta, P.C.: Phys. Rev. D **94**, 025031 (2016). arXiv:1604.01102 [hep-ph]
120. Rasmussen, C.Ø., Madsen, N., Robicheaux, F.: J. Phys. B: At. Mol. Opt. Phys. **50**, 184002 (2017); Corrigendum: J. Phys. B: At. Mol. Opt. Phys. **51**, 099501 (2018)
121. Malbrunot, C., et al.: Phil. Trans. Roy. Soc. Lond. A **376**, 20170273 (2018). arXiv:1710.03288 [physics.atom-ph]
122. Vargas, A.J.: Phil. Trans. Roy. Soc. Lond. A **376**, 20170276 (2018)
123. Ding, Y., Kostelecký, V.A.: Phys. Rev. D **94**, 056008 (2016). arXiv:1608.07868 [hep-ph]
124. Dehmelt, H., Mittleman, R., van Dyck, R.S., Jr., Schwinberg, P.: Phys. Rev. Lett. **83**, 4694 (1999). arXiv:hep-ph/9906262
125. Nagahama, H., et al.: Nature Commun. **8**, 14084 (2017)
126. Smorra, C., et al.: Nature **550**, 371 (2017)
127. Smorra, C., et al.: Nature **575**, 310 (2019)
128. Smorra, C., Mooser, A.: Precision measurements of the fundamental properties of the proton and antiproton. arXiv:2002.04261 [physics.atom-ph]
129. Bluhm, R., Kostelecky, V.A., Russell, N.: Phys. Rev. D **57**, 3932 (1998). arXiv:hep-ph/9809543
130. Testera, G., et al.: AEgIS collaboration. Hyperfine Interact. **233**, 13 (2015)
131. Pérez, P., et al.: GBAR collaboration. Hyperfine Interact. **233**, 21 (2015)
132. Bertsche, W.A.: Phil. Trans. R. Soc. A **376**, 20170265 (2018)
133. Müller, H., Hamilton, P., Zhmoginov, A., Robicheaux, F., Fajans, J., Wurtele, J.: Phys. Rev. Lett. **112**, 121102 (2014). arXiv:1308.1079 [physics.atom-ph]
134. Peters, A., Chung, K.Y., Chu, S.: Nature **400**, 849 (1999)
135. Charlton, M.: Phys. Lett. A **143**, 143 (1990)
136. Aghion, S., et al.: AEgIS Collaboration. Nature Commun. **5**, 4538 (2014)
137. Yu Voronin, A., Froelich, P., Nesvizohvsky, V.V.: Phys. Rev. A **83**, 032903 (2011)
138. Yu Voronin, A., Nesvizohvsky, V.V., Dufour, G., Debu, P., Lambrecht, A., Reynaud, S., Dalkarov, O.D., Kupriyanova, E.A., Froelich, P.: Int. J. Mod. Phys. Conf. Series **30**, 1460266 (2014)
139. Kostelecký, V.A.: Phys. Rev. D **69**, 105009 (2004). arXiv:hep-th/0312310
140. Kostelecký, V.A., Tasson, J.D.: Phys. Rev. Lett. **102**, 010402 (2009). arXiv:0810.1459 [gr-qc]
141. Kostelecký, A.V., Tasson, J.D.: Phys. Rev. D **83**, 016013 (2011). arXiv:1006.4106 [gr-qc]
142. Touboul, P., et al.: Phys. Rev. Lett. **119**, 231101 (2017). arXiv:1712.01176 [astro-ph.IM]
143. Hohensee, M.A., Estey, B., Hamilton, P., Zeilinger, A., Müller, H.: Phys. Rev. Lett. **108**, 230404 (2012). arXiv:1109.4887 [quant-ph]
144. Drummond, I.T.: Phys. Rev. D **88**, 025009 (2013). arXiv:1303.3126 [hep-th]
145. Drummond, I.T.: Phys. Rev. D **95**, 025006 (2017). arXiv:1603.09211 [hep-th]
146. Karshenboim, S.G.: Astron. Lett. **35**, 663 (2009). arXiv:0811.1008 [gr-qc]
147. Peters, A., Chung, K.Y., Chu, S.: Metrologia **38**, 25 (2001)
148. D. Stratikis *et al.*, FERMILAB-CONF-17-177-AD https://arxiv.org/ftp/arxiv/papers/1802/1802.C (2017)
149. Cox, S.F.J.: J. Phys C: Solid State Phys. **20**, 3187 (2017)
150. Hughes, V.W., Grosse Perdekamp, M., Kawall, D., Lui, W., Jungmann, K., zu Putlitz, G.: Phys. Rev. Lett. **87**, 111804 (2001)
151. Cassidy, D.B.: Eur. Phys. J. D **72**, 53 (2018)
152. Karshenboim, S.G.: Phys. Rep. **422**, 1 (2005)
153. Hayano, R.S., Hori, M., Horváth, D., Widmann, E.: Rep. Prog. Phys. **70**, 1995 (2007)

154. Hori, M., Walz, J.: Prog. Nucl. Part. Phys. **72**, 206 (2013)
155. Antipov, YuM, et al.: Sov. J. Nucl. Phys. **12**, 171 (1971)
156. Agakishiev, H., et al.: STAR collaboration. Nature **473**, 353 (2011). Erratum **475**, 412
157. Koch, H.: Hyperfine Interact. **44**, 59 (1988)
158. Johnson, C.D., Sherwood, T.R.: Hyperfine Interact. **44**, 65 (1988)
159. Cork, B., Lambertson, G.L., Piccioni, O., Wentzel, W.A.: Phys. Rev. **104**, 1193 (1956)
160. van Dyck, R.S., Schwinberg, P.B., Dehmelt, H.G.: Phys. Rev. Lett. **59**, 26 (1987)
161. Gabrielse, G., Fei, X., Orozco, L.A., Tjoekler, R.L., Haas, J., Kalinowsky, H., Trainor, T.A., Kells, W.: Phys. Rev. Lett. **65**, 1317 (1990)
162. Sellner, S., et al.: New J. Phys. **19**, 083203 (2017)
163. Ghosh, P.K.: Ion Traps. Clarendon Press, Oxford (1995)
164. Brown, L.S., Gabrielse, G.: Rev. Mod. Phys. **58**, 233 (1986)
165. Ulmer, S., *et al.*: Nature **524**, 196 (2015)
166. Gabrielse, G., Khabbaz, A., Hall, D.S., Heimann, C., Kalinowsky, H., Jhe, W.: Phys. Rev. Lett. **82**, 3198 (1999)
167. Yamazaki, T., Morita, N., Hayano, R.S., Widmann, E., Eades, J.: Phys. Rep. **366**, 183 (2002)
168. Hori, M., et al.: Science **354**, 610 (2016)
169. Korobov, V.I.: Phys. Rev. A **89**, 014501 (2014)
170. Hori, M., Sótér, A., Korobov, V.I.: Phys. Rev. A **89**, 042515 (2014). arXiv:1404.7819 [physics.atom-ph]
171. Keating, C.M., Charlton, M., Straton, J.C.: J. Phys. B: At. Mol. Opt. Phys. **47**, 225202 (2014)
172. Keating, C.M., Pak, K.Y., Straton, J.C.: J. Phys. B: At. Mol. Opt. Phys. **49**, 074002 (2016)
173. Pérez, P., Sacquin, Y.: Classical Quantum Gravity **29**, 184008 (2012)
174. Myers, E.G.: Phys. Rev. A **98**, 010101(R) (2018)
175. Zammit, M.C., et al.: Phys. Rev. A **100**, 042709 (2019)
176. van der Werf, D.P.: Int. J. Mod. Phys. Conf. Series **30**, 1460263 (2014)
177. Schiller, S., Balakov, D., Korobov, V.I.: Phys. Rev. Lett. **113**, 023004 (2014)
178. Amoretti, M., et al.: ATHENA Collaboration. Nature **419**, 456 (2002)
179. Gabrielse, G., et al.: ATRAP Collaboration. Phys. Rev. Lett. **89**, 213401 (2002)